M

Guide to

SBC05

Guide to

SBC05

JCT Standard Building Contract With
Quantities (SBC/Q)

JCT Standard Building Contract With
Approximate Quantities (SBC/AQ)

JCT Standard Building Contract
Without Quantities (SBC/XQ)

RIBA ⁂ Publishing

Sarah Lupton

Published by RIBA Publishing,
15 Bonhill Street, London EC2P 2EA

ISBN-10 1 85946 189 1
ISBN-13 978 1 85946 189 1

Stock code 56682

British Library Cataloguing in Publications Data
A catalogue record for this book is available from the British Library.

Publisher: Steven Cross
Commissioning Editor: Matthew Thompson
Project Editor: Anna Walters
Editor: David Hawthorn
Designed by Casciani Evans Wood
Typeset by Academic + Technical Typesetting, Bristol
Printed and bound by Hobbs The Printers, Hampshire
Figure 9 by Lucy Murawski

The text in Chapter 2 is adapted from an article by the Author which
first apeared in *International Construction Law Review*, Volume 23,
Part 1, January 2006, pp 90–101

Foreword

Long accepted as an industry standard, JCT98 and its predecessors had a tried and tested track record, lending it a reassuring familiarity to those whose job it was to administer it. The new edition, SBC05, with its more logical layout, clearer drafting style, and integration of many of the provisions previously contained in supplements, makes it easy to adopt and even easier to run. Changes in working practices, new legislation, and a large body of case law are all reflected in the new versions of the contract, which have been reduced from six to three by combining the local authorities and private editions.

Sarah Lupton's *Guide to SBC05*, which follows on from her excellent *Guide to JCT98*, is a straightforward and comprehensive analysis of the new form in the light of today's legal and practice landscape. Not only does she point out the important new changes, particularly to those provisions concerned with sub-letting terms, contractor's design obligations, extension of time, retentions, collateral warranties, third party rights and dispute resolution but she reflects on recent cases which serve as valuable reminders. The contract's provisions, procedures and supplementary conditions are spelled out knowledgeably and are organised by theme. The hard-pressed practitioner will be particularly pleased to see the useful indexes and clause comparison table, and will doubtless come to depend on being able to dip quickly into the book for specific help during the course of a job.

I would also thoroughly recommend the book to both architecture and other construction students on the threshold of undertaking their professional examinations. The comprehensive up-to-date coverage clearly and succinctly exposes the legal ramifications of the contract. Sarah Lupton's rare combination of being a legally-trained architect who also runs the MA in Professional Studies at Cardiff University makes this book the ideal student companion.

Professor Peter Hibberd
Secretary-General, The Joint Contracts Tribunal
June 2006

Contents

About the Guide

This is the second edition of the book originally entitled *Guide to JCT98*, first published in 1999. The original guide was intended as an introduction to the form for the student and newly qualified professional, and as a 'desktop' reference for the busy practitioner. The new guide maintains this approach, but has been revised to reflect the new edition of the standard form and recent changes in the law.

The guide does not assume any prior knowledge of either this form or the 1998 edition. Therefore the majority of the text does not refer to JCT98 or the changes brought about by the new edition of the form. However, for the benefit of those familiar with JCT98, chapter 2 comprises an analysis of the key differences between the forms, which are not confined to re-formatting and re-drafting but include some significant changes, such as the omission of the provisions for nominated sub-contractors and the inclusion of provisions for granting third party rights. As a further aid during the transition period, the guide contains two clause comparison tables, one of which has notes as to the more detailed changes. In addition to these changes the opportunity has been taken to clarify and expand parts of the text, and to add new case law and diagrams.

The guide gives a broad outline of the form and reasons why it might be selected, and examines the various documents that might form part of the contract package. Following Chapter 2 the guide contains further sections which examine the form on a topic by topic basis (rather than clause by clause), including timing and programming, control of the works (in particular, quality), payment, insurance, termination and dispute resolution. The guide also has a subject and a clause index.

This is the first book on which the author has worked without the invaluable advice and support of her long-standing colleague and friend, Stanley Cox, who sadly died last year. His wisdom, enthusiasm, and dedication to the profession are greatly missed by all those who knew him, and the author particularly misses the fascinating conversations over a mid-morning coffee at the Bute Building. This edition of the guide is dedicated to Stanley.

1 About SBC05

Key features

1.1 The JCT Standard Building Contract 2005 (SBC05) is the most recent edition of the long-standing JCT Standard Form of Building Contract. The current edition is published in three versions: With Quantities, Without Quantities and With Approximate Quantities. The differences between them relate to the documents on which the contract sum is based, and the calculation of sums due to the contractor. Unless otherwise indicated, this Guide makes reference throughout to the With Quantities version, with attention drawn to other versions where the differences are significant.

1.2 SBC05 is a traditional form of contract, and relatively simple in its overall structure. The contractor must carry out the work shown in the contract documents for the sum entered in the Contract Particulars, and within an agreed time period. Use of the form requires the appointment of an 'architect/contract administrator', a quantity surveyor and a planning supervisor, all of whom are named in the Articles. There are provisions for varying the work, and adjusting the contract sum and the completion date on the occurrence of certain events, all provisions shared with most other traditional standard forms of contract.

1.3 The contractor's primary obligation is to carry out the work shown or described in the contract documents, and the contractor takes overall responsibility for ensuring that the standards set out in the contract documents are achieved. In general terms, the form assumes that all work is designed by the employer's design team, and that the contractor will be supplied with all information necessary to carry out the Works. The contractor may, however, be required to carry out the design of a 'Contractor's Designed Portion'.

1.4 The provisions relating to the Contractor's Designed Portion in many ways reflect those of the JCT Design and Build Contract. The part of the Works to be designed is identified in the recitals, which also refer to the Employer's Requirements and the Contractor's Proposals for the Contractor's Designed Portion, and to the contractor's analysis of the portion of the Contract Sum relating to the Contractor's Designed Portion, termed the 'CDP analysis'. The Requirements are sent out with the tender documents and the Proposals returned with the tender, together with the analysis. The form includes a procedure for the submission of the related design information by the contractor for comment by the contract administrator, who retains responsibility for integrating the Contractor's Designed Portion with the rest of the design. The contractor is required to carry insurance to cover its design liability.

1.5 The form makes provision for two different methods of sub-contracting: to a domestic sub-contractor selected by the contractor and approved by the contract administrator; or to a sub-contractor chosen from a list of three named in the contract documents. With the contract administrator's approval the contractor's design obligations may be sub-contracted to a domestic sub-contractor. There are now two JCT sub-contracts specifically for use with SBC05: the Standard Building Sub-Contract, and the Standard Building Sub-Contract with Sub-Contractor's Design. There is no obligation in the main form for the contractor to sub-let on these terms, but there is a requirement for specific terms to be incorporated in any sub-contract.

1.6   The contract requires that the contractor commences work on an agreed 'date of possession', and completes the Works by an agreed 'date for completion'. The contractor is required to produce a master programme, although the form does not set out any particular requirements for the programme, or a sanction for its non-production. There are provisions which allow for the date for completion to be adjusted on the occurrence of specified events, and for the contractor to pay damages in the event of non-completion.

1.7   SBC05 allows for phased working, in that it is possible to divide the Works to be carried out into Sections, and set separate start and completion dates in relation to each Section. In general the conditions relating to timing operate independently with respect to each Section. For example, the contractor must notify the contract administrator of delays to any Section, and there is provision for fixing new completion dates for each Section as appropriate. A separate Practical Completion Certificate is required for each Section, and a separate Certificate of Making Good but only one Final Certificate. There are implications for non-completion, liquidated damages and retention.

1.8   Both the With Quantities and Without Quantities versions are lump sum contracts. In other words, all the work '… shown upon, described by or referred to …' in the contract documents must be carried out for the contract sum 'or such other sum as shall become payable hereunder at the times and in the manner specified in the Conditions'. The amount of work, which is covered by the contract sum, should be described in exact terms in the contract documents. In the With Quantities version the work is described in drawings and in a bill of quantities, whereas in the Without Quantities version the description is in drawings and a specification or schedules of work. Generally, if the description is inaccurate, any resulting addition to the cost is borne by the employer. If the contractor has made an error in pricing, however, then any shortfall will be borne by the contractor. The contract administrator has wide powers to order variations to the Works if required, and the contractor has a corresponding right to be paid any additional costs that arise from such variations.

1.9   JCT With Approximate Quantities is a re-measurement contract, where only approximate quantities are given for all of the work to be carried out, and the contract assumes that all work will be measured prior to certification. This version would normally be used where it may be difficult or impossible to measure the majority of the work accurately in advance, for example in a contract for refurbishment or repair following a fire or other damage. An approximate quantity may also be given for identified items when using SBC05 With Quantities. All versions allow for the use of provisional sums where it is impossible to specify or describe the work accurately in advance.

1.10   Payment to the contractor is made upon the issue of contract administrator's certificates at predetermined intervals. In general terms, the certificates will reflect the amount of work that has been properly completed up to the point of valuation in accordance with the terms of the contract. The provisions regarding payment, including the requirements for notices and the contractor's rights in the event of non-payment, reflect those required under the Housing Grants, Construction and Regeneration Act 1996 (see Figure 1).

1.11   The contract administrator has a significant role under the contract, which includes the issuing of certificates and the power to order variations to the Works. At some points the contract administrator is acting as the employer's agent, and at others as an independent administrator. A court would assume that the parties have contracted on the basis that the contract administrator will act fairly at all times in applying the terms of the contract and particularly so when acting as an independent administrator. This duty of fairness, however, does not place the contract administrator in the same position as an arbitrator, in that the contract administrator is not immune from being sued.

1.12   In addition to the core provisions concerning quality, design, programme and payment, SBC05 also contains detailed provisions covering matters such as injury, damage and insurance, third party rights and warranties, default and termination, and dispute resolution, all of which are discussed in detail in this Guide.

Figure 1 Provisions required under the HGCR Act 1996

HGCR Act 1996	SBC05 clause	Provision concerning
S108	Article 7, cl 9·2	Adjudication
S109	Cl 4·9	Stage payments
S110 (1)	Cl 4·9, 4·11, 4·13·1, 4·15·1, 4·15·4	Dates for payment
S110 (2)	Cl 4·13·3, 4·15·3	Notices
S111 (1)	Cl 4·13·5, 4·15·5	Notices, payment
S111 (1), (2)	Cl 4·13·4, 4·15·4	Notices
S112	Cl 4·14	Contractors right of suspension

Deciding on SBC05

1.13   The JCT Standard Form of Building Contract has been the first choice of form for many contemplating a traditional procurement route. Long accepted as an industry standard, it has a tried and tested track record that gives users the reassurance that there should be little to surprise them, and much guidance available on its use. In particular the new edition, with its more logical layout, clearer drafting style, and integration of many of the provisions previously contained in supplements, should be an attractive option to those already familiar with the form (see Chapter 2 for detailed discussion of the changes). However, there are some aspects that should be given careful consideration before deciding to proceed with this form.

1.14 Compared with some other traditional forms such as the JCT Major Project Construction Contract (MPF), this form places more risk on the employer. One of the primary functions of any construction contract is to allocate risk between the parties. In many ways it is meaningless to describe any particular contract as being 'fair' or 'unfair' in that, as long as the allocation of risk reflects what the parties intended, they enter into it as a commercial decision and the price agreed would reflect that balance. The more risk the employer is prepared to accept, the lower the price is likely to be. A good example from SBC05 is the operation of the fluctuations clauses, which allow the contract sum to be adjusted to take account of changes in market prices. If this provision is not used then the contractor must anticipate the risk of price changes, and this may result in a relatively high tender.

1.15 A further example is that the responsibility for providing the contractor with all information reasonably necessary to construct the project lies with the employer, except for work forming part of the Designed Portion. Even here, the contractor is not responsible for checking the employer's requirements, and is entitled to proceed with its design on the assumption that all information provided is accurate. Any delay in providing information, or in commenting on information provided by the contractor, is the employer's risk. The contractor is also entitled to claim additional time, upon the occurrence of various neutral events, such as exceptionally adverse weather. It is therefore important that anyone advising an employer on the possible selection of SBC05 as the contract form to be used on a project understands how the balance of risk between the parties is affected.

1.16 SBC05 can admittedly be criticised for being lengthy and complex in its procedural rules. The construction of large projects is, however, a complex process often involving many risks, therefore in order for parties to decide in advance what should be the outcome of various eventualities the form is necessarily detailed and long. By comparison it should perhaps be pointed out that there are far longer forms of building contract in regular use today, for example the Engineering Advancement Association of Japan (ENAA) Model Form of Contract runs to five volumes! An alternative approach, and that adopted by MPC, is to provide the parties with a set of core clauses, and require them to draft additional terms to suit their particular needs, but this method requires expert advice, and is likely to produce a document at least as long as SBC05, or otherwise the parties will be left 'in the dark' as to their relative positions in certain circumstances. Administrators of large and complex projects must be prepared to undertake the sophisticated procedures and levels of complexity that attend major forms of contract.

1.17 JCT Ltd has had to respond to changes in working practices, to take account of new legislation, and to act where the judiciary have indicated that parts of the form are unclear, or have sometimes placed an interpretation on the form that it was never intended to have. As a result of this the provisions of SBC05 are balanced and finely honed. To carry out ad-hoc amendments can produce an imbalance and bring unexpected consequences. The new edition, in particular, has been adjusted to respond to working procedures that

Figure 2 Comparison of provisions

	MP	SBC	IFD	IF	MWD	MW
Contractor design	yes	yes	yes		yes	
'Contractor's Design Submission Procedure'	yes	yes				
Possession by sections		yes	yes	yes		
Completion by sections	yes	yes	yes	yes		
Deferment of possession		yes	yes	yes		
Information release schedule		yes	yes	yes		
Partial possession	yes	yes	yes	yes		
Employer's representative	yes	yes				
Clerk of works		yes	yes	yes		
Listed sub-contractors		yes				
Named sub-contractors/specialists	yes		yes	yes		
Pre-appointed consultants	yes					
Advance payment, bond		yes	yes	yes		
Activity schedule		yes	yes	yes		
Application by contractor	yes	yes	yes	yes		
Payment for off-site materials, bond		yes	yes	yes		
Retention bond		yes				
Variation quotations	yes	yes				
Interest on late payment	yes	yes	yes	yes	yes	yes
Professional indemnity insurance	yes	yes	yes			
Joint Fire Code		yes	yes	yes		
Third party rights	yes	yes				
mediation	yes	yes	yes	yes	yes	yes
adjudication	yes	yes	yes	yes	yes	yes
arbitration		yes	yes	yes	yes	yes
litigation	yes	yes	yes	yes	yes	yes
Code of procedure for testing		yes				
Fluctuations options		3	1	1	1	1

have developed over the last few years, and to incorporate several of the ad-hoc amendments frequently made in practice.

1.18   A significant change under the new edition is the removal of the nominated sub-contracting provisions. The Guide to SBC05 states that this is to reflect industry trends, but the removal leaves the employer without any obvious means of requiring the contractor to use any particular sub-contractor. If the SBC05 'listing' provisions are considered not sufficiently prescriptive, then another form such as either the MPF with its provisions for 'Named Specialists' or the Intermediate Building Contract, with its naming provision, may have to be considered as an alternative.

1.19   SBC05 nevertheless contains some features that are unique among JCT forms for traditional procurement, including the procedure for 'pre-agreed adjustments', the provisions for 'third party rights', and the option of a bond in lieu of retention (see Figure 2). In situations where these features are an advantage, and the ability to 'nominate' is not a requirement, SBC05 will no doubt remain a popular choice.

2 Comparison with JCT98

2.1 The new Standard Building Contract is essentially the same contract as the previous edition (JCT98), but the differences are not restricted to simply a re-packaging and re-numbering of the existing clauses. Figure 3 shows the more significant changes, most of which are discussed below. A complete clause conversion chart is included in Appendix A.

General changes

2.2 The number of variants of the form has been reduced from six (for JCT98) to three (for SBC05). The reduction is achieved by combining the 'Local Authority' and 'Private' editions – the differences between the two had in any case been minor, and are now allowed for by options within the form. The Contractor's Designed Portion and Sectional Completion Supplements have been integrated with the form, which makes using these provisions much easier than the laborious process of trying to read the former schedules of amendments alongside the main form, as had been the case with JCT98. Together with the incorporation of the fluctuations provisions and the new 'third party rights' schedules, this change reduces the proliferation of ancillary documents.

2.3 The most obvious changes are to appearance, layout, content and style, changes that have been applied to the full JCT suite of forms. The structure and clause numbering have been adjusted so that all the form is now in 'section-headed format'. This, coupled with the inclusion of the specific requirements of the parties under a 'Contract Particulars' section at the front of the form (in place of the former 'Appendix'), makes the form much easier to navigate. The language has been clarified throughout, by the removal of redundant phrases, the breaking down of overly long sentences and paragraphs, and the addition of several new 'defined terms' (such as 'employer's persons'), which removes the need to repeat such definitions within the clause itself. The result of this is that several sections of the form, in particular the termination and insurance provisions, have been significantly reduced in length. The author was pleased to see (finally!) a new clarifying clause dealing with gender bias (clause 1·4·3).

2.4 Admittedly there are some places where this could have been carried one step further. For example, under the Contract Particulars the reference to 1·8 now refers to all communications being in writing, unless it has been agreed that they may be made electronically, yet the requirement for a notice, instruction, etc., to be in writing is repeated throughout the form. Also, no subject index is included, which might have been a great help to first time users (and experienced users until they get used to the new layout). The Guide to the form includes a very useful 'table of destination', which shows where JCT98 clauses appear in SBC05, but unfortunately not an equivalent 'table of origins'; therefore it is not immediately apparent which clauses are entirely new. Nevertheless, despite these minor criticisms, the form is much clearer and more user-friendly than its predecessor.

Figure 3 Key changes

Layout/terminology differences:
- Contract Particulars follow Articles, replacing the Appendix.
- '13A Quotations' under a schedule.
- Insurance under a schedule.
- Architect becomes Architect/contract administrator.
- 'Employer's Persons' and 'Contractor's Persons' defined.
- 'Insolvent', 'Purchaser and Tenant' and 'Relevant Omissions' defined.
- Extension of time section now headed 'Adjustment of completion date'.
- Defects liability period now 'Rectification period'.
- Determination now 'termination'.

Omissions:
- Nominated sub-contractor provisions removed.
- Nominated supplier provisions removed.
- Performance specified work provisions removed.
- Contractor's Price Statements removed ('13A' retained).
- Clause 22D insurance omitted (loss of liquidated damages).
- Construction Industry Scheme provisions largely removed.
- VAT supplementary agreement removed.
- Requirement to certify 'frost damage' under clause 17·5 removed.

Additions:
- Contractor's Designed Portion Supplement provisions now integrated.
- CDP PII insurance requirement added.
- CDP design submission procedure added.
- Copyright in the Contractor's Design Documents added.
- Sectional Completion Supplement provisions now integrated.
- Fluctuations options included (as a schedule).
- 'Third party rights' included as a schedule (Purchaser/Tenants, Funders).
- Inclusion of provisions for collateral warranties from the contractor (Purchaser/Tenants, Funders: the warranties themselves are separate documents).
- Inclusion of provisions for collateral warranties from sub-contractors (Purchaser/Tenants, Funders and Employer: the warranties themselves are separate documents).

Other changes:
- EDI replaced by broader 'electronic communications' provision whereby parties may agree their own procedure.
- Extension of time provisions revised, relevant events reduced, CA to notify contractor 'as soon as reasonably practicable', decision to be apportioned.
- CA to give reasons for dissatisfaction with work (cl 3·20).
- Default retention deduction reduced from 5 per cent to 3 per cent.
- Determination (now 'Termination') provisions simplified.
- Arbitration – default changed to legal proceedings.
- Adjudication now under the Scheme.
- Mediation moved from footnote to an express clause.
- English courts to have jurisdiction over any dispute.

Key changes

Sub-contracting

2.5   The most significant change to content is undoubtedly the removal of the nominated sub-
        contractor provisions, and indeed any reference to these. The reason given by the JCT is
        as follows:

> there appears to have been little use of the provisions for Nominated Sub-contractors
> and little appropriate use of the Nominated Supplier Provisions. The provisions of
> JCT98 for listing of sub-contractors have been retained and, in the JCT's view, the
> specifying of a supplier is a matter generally better dealt with in the Contract Bills or
> other Contract Documents. (SBC05 Guide, page 1)

2.6   Employers wishing to 'nominate' a range of sub-contractors may now have to consider
        using a different procurement route (for example the Construction Management family of
        forms). Alternatively, if they wish to continue using traditional procurement they could
        consider the Major Project Form, where the employer may pre-select specialists for whom
        the contractor remains entirely responsible, or the Intermediate Building Contract, which
        has retained its naming provisions. A third option would be to specify a sub-contractor in
        the Bills of Quantities. The latter route is, of course, problematic, unless carefully drafted
        clauses are included to make it clear where liability will lie for any default of the specified
        sub-contractor. There has been conflicting case law relating to the main contractor's
        responsibility, and it is at least arguable that without amending the form the main contractor
        may not be responsible for any design carried out by the apecified sub-contractor, for
        finding another sub-contractor should it repudiate the contract, or for remedying its defective
        work.

Sub-letting terms

2.7   The sub-letting provisions have been altered to include two additional requirements; any
        sub-contract must provide for the execution and delivery by the sub-contractor of warranties
        required under the contract documents, and must require the sub-contractors to allow
        access to workshops and off-site work (clause 3·9).

Contractor's obligations

2.8   The contractor's primary obligation remains 'to carry out and complete the Works' (clause
        2·1, in both JCT98 and SBC05). At first sight it appears that the JCT have dropped the
        qualifying statement 'provided where and to the extent that approval of the quality of
        materials or of the standard of workmanship is a matter for the opinion of the Architect,
        such quality and standards shall be to the reasonable satisfaction of the Architect', but the
        phrase has been amalgamated with text from what was formerly clause 8·1·2, to form a
        combined clause 2·3·3 which states:

> Where and to the extent that approval of the quality of materials or goods or of
> the standards of workmanship is a matter for the opinion of the Architect/Contract
> Administrator, such quality and standards shall be to his reasonable satisfaction. To
> the extent that the quality of materials and goods or standards of workmanship are

neither described… nor stated to be a matter for such opinion or satisfaction, they shall in the case of the Contractor's Designed Portion be of a standard appropriate to it and shall in any other case be a standard appropriate to the Works.

2.9   Under the old wording the requirement to achieve a standard 'appropriate to the Works', in the absence of any specified standard, applied only to workmanship. As a consequence of the above change this default now applies also to materials and goods. In many instances the obligation to provide something 'fit for purpose' might have been implied, but as a result of the change it will no longer be necessary to argue for an *implied term* in cases where, for example, the specification of the particular material or good is (inadvertently or deliberately) incomplete, an argument which has not always succeeded. It is not clear whether or not this change was intended by the JCT, as no reference to it is made in the Guide.

Contractor design

2.10   The integration of the Contractor's Designed Portion Supplement allows for the contractor to be required to design an identified part of the Works. (This is effected by the seventh recital extending the definition of the 'Works' which the contractor must carry out and complete to include 'the design and construction of the Contractor's Designed Portion'.) The form anticipates that, depending on the information requested at tender stage, the contractor's design proposal may not be fully detailed before work commences, and contains detailed provisions regarding the submission of the developing design by the contractor, for comment by the contract administrator (the Schedule 1 provisions). These provisions give the contract administrator a reasonable degree of control over this aspect of the contract.

Insurance

2.11   The contractor is now required to carry professional indemnity insurance. The amount of cover and the period of expiry are inserted in the Contract Particulars. The insurance must be taken out immediately following execution of the contract. There is provision for inserting a level of cover for pollution or contamination claims – if none is inserted, the level is the same as the level of cover inserted in the Contract Particulars.

2.12   The insurance provisions otherwise remain largely unaltered. Minor changes include the removal of the clause 22D provisions (employer's insurance for loss of liquidated damages) and the deletion of the limitation to damage caused by 'fire and explosion' in the definition of Terrorism Cover.

Timing

2.13   The broad system with regard to commencement, progress and completion remains the same. The integration of the sectional completion provisions means that the work may split into sections, each with its own start and finish date. All the provisions relating to timing (for example the obligation to issue a Non-completion Certificate) apply separately to each section, except that there is only one Final Certificate.

2.14   The heading 'Extension of time' has been replaced with 'Adjustment of completion date' to reflect the facility under Schedule 2 (formerly clause 13A) for agreeing acceleration to the contract. The clauses, however, still refer to the process of giving an extension of time. Giving a notice with particulars, including an estimate of the expected delay, remains a condition precedent for the award of an extension of time (cl 2·28·1). The contract administrator must now respond to the notice 'as soon as is reasonably practicable' and not later than 12 weeks (cl 2·28·2). Clause 2·28·3·1 requires the contract administrator to break down any extension by apportioning the time to each Relevant Event and Relevant Omission (this replaces clauses 25·3·2 and 26·3, i.e. the obligation is no longer limited to those events which relate to loss and/or expense claims). Following any notification by the contractor of a delay, clause 2·27·3 now states 'the contractor shall forthwith notify the Architect/Contract administrator in writing of any material change in the estimated delay'. The qualification of 'reasonably necessary' in JCT98 clause 25·2·3 has been dropped and the obligation to keep the architect informed is now strict.

2.15   The list of Relevant Events has been considerably shortened. This was pre-empted by Amendment 4 (January 2002): Extension of Time/Loss and Expense/Advance Payment to JCT98, which introduced an important new Relevant Event/Matter, in the form of a 'sweeping up' clause covering any default or omission of the employer. This is now incorporated in clause 2·29 of SBC05. As the general event duplicated several of the JCT98 more specific events, these have now been dropped, namely:

- failure of the contract administrator to supply information;

- execution of work not forming part of contract by employer;

- supply of materials and goods by employer;

- failure to provide ingress and egress to the site;

- failure to comply with CDM obligations.

2.16   More importantly, 'contractor's inability for reasons beyond his control and which he could not reasonably have foreseen... to secure labour' (or goods/materials) has been removed as a Relevant Event, reflecting the general trend in practice to delete this event.

Payment

2.17   In general the provisions regarding payment remain unaltered. In respect of pricing variations, the '13A quotation' provisions have been retained (now termed a Schedule 2 quotation), whereas the provisions allowing the contractor to submit a 'price statement' have been dropped. The difference between the two systems had been that whereas under a '13A quotation' the contractor must submit a price (and an extension of time, and loss and expense assessment) if requested to do so, and should not execute the variation until the price is agreed, in the latter system the contractor could submit a price on its own initiative, and must execute the variation. It would still be open to the contractor to submit its own claim even if not asked, but the QS would no longer be obliged under the contract to give reasons for rejecting that assessment, as it had under the 'price statement' system.

2.18   The default retention has now been changed to 3 per cent from 5 (clause 4·20·1). Where a bond is to be provided in lieu of retention, it is now expressly provided that this bond, in common with the Advance Payment Bond and the Bond in respect of payment for off-site goods and materials, should be from a surety provided by the Employer. Other points to note are that clause 30·1·1·5 has been re-worded (under clause 4·13·5) so that the amount to be paid if no 'withholding notice' is issued is the amount shown in the Interim Certificate (and not 'the amount due pursuant to clause 30·1·1·10'), reflecting the equivalent provision for the Final Certificate (30·8·4).

Warranties/third party rights

2.19   SBC05 allows for requiring the contractor to provide collateral warranties to a funder, or to purchasers/tenants, and in addition for the contractor to obtain warranties from its sub-contractors to either of these or to the employer. The JCT intends to publish a range of standard forms of warranty to suit all these situations (SCWa/F, SCWa/P&T, SCWa/E). In respect of warranties to funders or purchasers/tenants SBC05 states that the JCT forms will be used. In relation to warranties from sub-contractor to employer, the form requires that the form of warranty is specified in the Contract Documents. (The reason for this difference appears to relate to the publishing timetable, as the Guide indicates that SCWa/E will be incorporated once this form has been produced.) In all cases the details of the intended beneficiaries who may be identified by name, as a member of a class, or as matching a description, are to be set out in Part 2 of the Contract Particulars.

2.20   As an alternative to collateral warranties, the form contains new provisions for granting 'third party rights', relying on the facility offered by the Contracts (Rights of Third Parties) Act 1999. The beneficiaries are identified in Part 2 of the Contract Particulars, and can be specifically named, as a member of a class, or as matching a description. The rights which the contractor agrees to grant are set out in Schedule 5 and are largely similar to those set out in the equivalent collateral warranties.

Termination

2.21   The provisions for determination (now 'termination') have been considerably abbreviated, largely by grouping some of the consequences of termination under a separate section, rather than repeating them for both employer and/or contractor initiated termination. The provision for automatic termination in the event of the contractor's insolvency or bankruptcy has been removed. References to 'a 17·5·2·1 agreement' (i.e. continuation or novation agreements) and to 'interim arrangements' have also been dropped. These were in effect enabling provisions, and the Guide suggest that the employer's insolvency practitioner will advise on appropriate agreements and arrangements where a rescue procedure may be a practical proposition.

Dispute resolution

2.22   There have been some changes to the dispute resolution provisions. The reference to mediation, which was formerly in a footnote, has now been incorporated as a clause in the form, but remains an enabling provision, as there is no requirement to engage in a mediation process. There are also no details as to the method of appointing a mediator or the procedure to be followed. In relation to the statutory right to adjudication, the JCT provisions for adjudication have been almost entirely dropped, and the procedure set out in the secondary legislation 'the Scheme for Construction Contracts' will now apply. The provisions relating to the naming and appointment of an adjudicator have been retained, as have those relating to cases of opening up and testing.

2.23   The arbitration provisions are largely unaltered, and refer to the JCT 2005 edition of the CIMAR rules. A significant alteration is that litigation has replaced arbitration as the default method of dispute resolution (subject, of course, to the right to adjudication). Those wishing to use arbitration must therefore not forget to state in the Contract Particulars that 'Article 8 and clauses 9·3 to 9·8 (Arbitration) apply' (under article 8 the parties agree that all disputes arising out of or in connection with the contract shall be referred to arbitration). With respect to litigation, those who use the form for work with an international element should note that there is now an article 9, which states that 'the English courts shall have jurisdiction over any dispute or difference between the Parties which arises out of or in connection with this contract'.

3　The contract documents

3.1　A contract entered into on the basis of SBC05 will comprise an extensive package of documents, the majority of which will have been issued to the contractor at tender stage. Documents are central to the success of every building operation, and in traditional procurement in particular the contractor depends on full and accurate information being provided in adequate time and to a pre-determined pattern. Generally firm and full information at tender stage reduces the risk of cost increases and programme alterations later in the contract.

3.2　When using SBC05, the primary document is of course the printed form itself, which comprises not only articles and Conditions but also various schedules, which include 'third party rights', forms of bonds and fluctuations provisions (see Figure 4 below). SBC05 also makes reference to various other documents. Some of these are termed 'Contract Documents' (see below) whereas others are referred to at various places in the form, for

Figure 4　Layout of the form

Articles of Agreement:
Recitals
Articles
Contract Particulars
Attestation

Conditions:
Section 1: Definitions and Interpretation
Section 2: Carrying out of the Works
Section 3: Control of the Works
Section 4: Payment
Section 5: Variations
Section 6: Injury, Damage and Insurance
Section 7: Assignment, Third Party Rights and Collateral Warranties
Section 8: Termination
Section 9: Settlement of Disputes

Schedules:
Schedule 1: Contractor's Design Submission Procedure
Schedule 2: Schedule 2 quotations (replaces 13A)
Schedule 3: Insurance Options
Schedule 4: Code of Practice (for opening up and tests)
Schedule 5: Third Party Rights
Schedule 6: Forms of Bonds
Schedule 7: Fluctuations Options

example the recitals refer to an Activity Schedule and an Information Release Schedule, and the articles refer to the Construction Industry Model Arbitration Rules – all significant documents. Although not termed 'Contract Documents', many of these may form part of the contract between the parties. Indeed any document to which clear reference is made in the 'Contract Documents' will form part of the binding agreement between the parties.

3.3   The documents used will, to a certain extent, depend upon the version of SBC05 that is selected. In addition, if the Works are to include a Contractor's Designed Portion, this will affect the documents to be used. Figure 5 indicates some of the possible combinations of documents that may make up the contract package.

Figure 5 Documents

	With Quantities	With Approximate Quantities	Without Quantities, Alternative A	Without Quantities, Alternative B	CDP
Drawings	CD	CD	CD	CD	
Bills	CD	CD			
Priced specification or Work Schedules			CD		
CDP Analysis or Schedule of Rates				CD	
Specification				CD	
Rule 116 Schedule (Fluctuations Option C only)			CD	CD	
Priced Activity Schedule (optional)	R		R	R	
Information Release Schedule (optional)	R	R	R	R	
Employer's Requirements					CD
Contractor's Proposals					CD
Contractor's Designed Portion Analysis					CD
'the Listed Items' and related Bond	cl 4·17	cl 4·17	cl 4·17	cl 4·17	
Advance Payment Bond	cl 4·8	cl 4·8	cl 4·8	cl 4·8	
Retention Bond	cl 4·19	cl 4·19	cl 4·19	cl 4·19	
Health & Safety Plan	cl 3·25	cl 3·25	cl 3·25	cl 3·25	

KEY
CD = termed a 'Contract Document'
R = referred to in the recitals
cl = referred to in the conditions

'Contract Documents'

3.4   SBC05 With Quantities defines 'Contract Documents' as 'the Contract Drawings, the Contract Bills, the Agreement and these Conditions together with (where applicable) the Employer's Requirements, the Contractor's Proposals and the CDP Analysis' (cl 1·1). The Agreement and Conditions are, of course, found in the form. SBC05 With Approximate Quantities also includes contract bills within the definition of 'Contract Documents', although in this case all the quantities will be approximate. SBC05 Without Quantities defines Contract Documents as including '(where Pricing Option A applies) the Priced Document or (where Pricing Option B applies) the Specification.' The former is where the contractor has supplied a priced specification or Works Schedules (the Priced Documents). The latter is where the contractor has stated a lump sum only and is required in addition to supply either a Contract Sum Analysis or a Schedule of Rates, which is referred to as 'the Priced Document', but is not defined as a Contract Document. In both cases, where Fluctuations Option C applies, the Schedule required by rule 11b of the JCT Formula Rules will also be a Contract Document.

3.5   The articles of agreement and the Contract Particulars must be completed very carefully. The articles of agreement contain the attestation that must be signed by both parties and witnessed. This forms the heart of the agreement whereby the contractor undertakes to 'carry out and complete the Works in accordance with the Contract Documents' (article 1), and in return the employer undertakes to pay the contractor the contract sum as adjusted in accordance with the Conditions (article 2).

Contract drawings

3.6   The 'Contract Drawings' are listed under the third recital (or the second recital in the Without Quantities version). These should all be identified precisely, including revision numbers, etc. The list may be annexed if long, but if so the list must be clearly identified. Note that in SBC05 there is no reference to who prepared the drawings.

Contract bills

3.7   The 'Contract Bills', unless otherwise stated, must be prepared in accordance with the Standard Method of Measurement of Building Works, 7th edition (SMM7, cl 2·13·1). If materials or goods are to be paid for prior to delivery on site, a list of these must be annexed to the bills (cl 1·1, definition of 'Listed Items'). Bills of quantities are normally prepared by the quantity surveyor based on detailed drawings and a specification prepared by the contract administrator. The use of bills does not reduce the responsibility of the contract administrator for the preparation of that information. For good practice in preparation and coordination of specification, drawings and bills of quantities see current relevant publications on Coordinated Project Information (CPI). The contract requires that the parties sign both the drawings and the bills (second and third recitals).

3.8   In practice, even where bills of quantities are used, a specification often forms part of the documentation, either bound in as a section of the bills, perhaps as part of the preambles, or as a separate document referred to in the bills. The CPI recommendation is that the specification becomes the core document in terms of defining quality, and that the drawings and the bills refer to clauses in the specification. If this system is used, the specification will be an essential part of the package and it is suggested that it should be signed with the other 'Contract Documents'.

3.9   The third recital (Alternative A) of the Without Quantities edition refers to 'priced Specification or Works Schedules'. One of these should be deleted in the contract particulars as appropriate, though there is no reason why both a specification and schedules should not be used in the contract package. The contract does not define what form the schedules should take but the documents could be arranged by work Sections, by trades or, as is frequently used in refurbishment, on a 'room-by-room' basis. In all cases it is likely to be clearer if the detailed specification information is kept in a separate document referred to by the schedules. If this is done then it is suggested that, as the contractor would normally be asked to price the schedules, the reference to the priced specification is deleted in the contract particulars. The specification should then be bound into or identified and referred to in the schedules, and signed with the other contract documents.

Employer's Requirements

3.10   The 'Employer's Requirements' are referred to in the eighth recital as 'documents showing and describing or otherwise stating his requirements for the design and construction of the Contractor's Designed Portion', and the form assumes that these have been sent to the contractor at tender stage.

3.11   The contract does not stipulate any format for the Employer's Requirements. In broad terms, the documents will set out the employer's requirements for the Contractor's Designed Portion of the Works. Given that clause 2·2 requires the contractor to complete the design 'in accordance with the Contract Bills' rather than the Employer's Requirements, it would be sensible to include the Requirements within the bills, or make a clear reference to them within the bills. The Requirements should be prepared carefully and on the assumption that there will be no changes to the Requirements once the contract is let, for although the contract contains provisions whereby a variation can be instructed, such variations may result in additional costs to the employer, and are subject to the consent of the contractor.

3.12   The Requirements could be in a very summary format, for example simply giving a brief description of the relevant part or system, referencing to drawings indicating its location and coordinating dimensions. It is likely, though, that they will be more detailed than that and include a detailed specification, in either prescriptive or performance terms, or in all probability involving a mixture of the two. They could also include schematic layouts or outline designs of the relevant part. In essence they act as a brief. Advice on briefing is outside the scope of this Guide but reference could be made, for example, to David Hyams, *Construction Companion to Briefing*, RIBA Publications (2001). Consultants may also find *The JCT Guide to the Use of Performance Specifications*, JCT (2001) very helpful when preparing the Requirements.

3.13   One of the most important inclusions is to stipulate in exactly what form the Proposals should be submitted, and what they should include. This is essential in order for the employer to make a clear assessment of the submitted tenders. The amount and level of detail of the information will depend upon the scale of the Designed Portion, and its relationship with the rest of the design. Where the Contractor's Designed Portion forms a significant element in the project, full information may be needed in order to integrate this element with other elements of the design, and in such cases the employer may need to adopt a two-stage tender approach.

3.14   It is also very important that the Requirements should specify the drawings and other design information (the 'Design Documents') to be submitted by the contractor following acceptance of tender, and a programme for their submission. The purpose of this is to control the scope, format and timing of the submission of Design Documents for review. For example, it should protect the contract administrator from being overwhelmed by Design Documents at an inconvenient time, or from being presented with Design Documents to review for key elements in isolation from information on other related aspects of the design. It is likely that the programme will be the subject of negotiation at tender, as it is important that any programme in the Requirements will also meet the contractor's needs in terms of developing the design at a rate which will support its intended construction programme. It would also be wise to set out the information required to be submitted at practical completion, such as 'as built drawings', otherwise the contractor's obligation is to provide such information 'as the employer may reasonably require' (cl 2·40).

Contractor's Proposals

3.15   The Contractor's Proposals should be in the format and contain the information stipulated in the Employer's Requirements. These may request that various documents are provided, including drawings, specifications, schedules, programmes, method statements, etc.

3.16   The contractor should raise matters relating to the contract data where decisions are outstanding from the employer, so that these can be resolved. The Proposals should indicate clearly any areas of conflict in the Requirements, and any instances where the contractor has found it necessary to amend or amplify the brief. The contract does not allow for the inclusion of provisional sums in the Proposals, only in the Requirements, so if the contractor wishes to cover any part of the Proposals with a provisional sum then it should inform the employer so that the Requirements can be amended.

CDP Analysis

3.17   The contract does not proscribe a format for the CDP Analysis. It would therefore be sensible to set out what format would be acceptable in the Employer's Requirements. (It would not be unreasonable, in cases where the Designed Portion forms a significant part of the works, for the contractor to be asked to prepare a full bill of quantities, although this would be unlikely on smaller projects.) The contract requires that the document is used for assessing the value of employer-instructed variations to the Contractor's Designed Portion (cl 5·8·2). It does not require that the document is used to assess the value of work carried out, etc., to be included in periodic payments, but it would normally be used by the contractor to prepare applications for payment, and by the employer in checking such applications.

Other documents

Activity Schedule

3.18 The second recital (except for the With Approximate Quantities version) refers to a priced Activity Schedule, a provision that may be deleted if not required. The Schedule is prepared and priced by the contractor and provided prior to the contract being executed. An example of a priced Activity Schedule was included in the guidance notes to JCT80 Amendment 18, and it is very similar to a schedule of work (see Appendix C at the back of this book). Each activity is priced, and the sum of those prices will be the contract sum with certain exclusions such as provisional and prime cost sums (see footnote [3] to the second recital). If it is included, it is used to ascertain the value of work properly executed for certification purposes (cl 4·16·1·1).

3.19 An Activity Schedule is unlikely to be useful where full 'Work Schedules' have been prepared by the design team, unless they have been prepared on a different format to that suggested for the Activity Schedule, for example on a 'room-by-room' basis. It may be useful alongside bills of quantities if it saves time, and therefore consultant's fees, at the time of Interim Certificates, but may result in a loss of accuracy of valuations. It is likely to be most helpful where the tender package consists of drawings and specification alone.

Information Release Schedule

3.20 The Information Release Schedule, referred to in the fifth recital is an optional provision (the Schedule is deleted if it is not required). The Schedule should state 'what information the Contract Administrator will release and the time of that release'. If used, the Schedule should be prepared by the contract administrator and sent out with the tender documents. The Schedule does not need to list all the information that will be provided but should, for example, list key drawings.

3.21 The Schedule will of course make it clear to the contractor in advance when information will be provided, and will therefore enable the contractor to programme the work more effectively, and possibly reduce the number of potential arguments that may arise regarding delays. A significant implication for the contract administrator is that if any information listed is provided later than the stipulated date then this will be a Relevant Event in relation to an extension of time. It should also be noted that if there is any adjustment to the completion date then adjustments to the Schedule might have to be negotiated between the parties.

Health and safety documents

3.22 Where all the Construction (Design and Management) Regulations 1994 (CDM Regulations) apply to the particular contract, as will usually be the case, clause 3·25 specifically refers to the Health and Safety Plan, and the Health and Safety File.

3.23 The Health and Safety Plan is not a contract document under SBC05, and the recitals make no mention of it having been prepared and given to the contractor at the time of tender. Nevertheless, it is a statutory obligation for the employer to have had one prepared and passed to the principal contractor under Regulation 15 of the Construction (Design and Management) Regulations 1994. As it may have pricing and timing implications a 'pre-tender' Health and Safety Plan is usually sent out with the tender documents. Before construction work can begin the contractor must have developed the plan to comply with Regulation 15(4). To avoid uncertainty, it is advisable to require that this document be submitted well in advance of the Date for Possession.

3.24   The Health and Safety File is principally a matter for the planning supervisor who will compile it but there is a requirement on the contractor to provide information for this File, and to ensure any sub-contractor also complies. The contract administrator, when certifying practical completion, must make sure that the contractor has 'sufficiently complied' with this requirement before issuing the certificate.

Bonds

3.25   SBC05 includes three forms of bond: (1) an advance payment bond, (2) a bond in respect of payment for off-site materials and/or goods and (3) a bond in lieu of retention. Where required, the contractor must arrange bonds, and as all of these are optional it must be made clear to the contractor at tender stage if any will be required. An Advance Payment Bond is normally required where an advance payment is to be made to the contractor under clause 4·8 (note that this is not an option where the employer is a local authority). A bond in respect of payment for off-site materials and/or goods is required where it has been agreed that certain materials or goods will be paid for in advance of them being brought on site – the so-termed 'listed items' (cl 4·17). A bond in lieu of retention is an alternative form of security to the more traditional use of a retention deduction. Terms for all of the bonds have been agreed between the British Bankers Association and JCT Ltd, and are included in the form under Schedule 6. If any other terms are preferred, or if any other type of bond is required, for example a performance bond, then these must be available to the contractor before the contract is entered into.

Sub-contract documents

3.26   JCT Ltd publish two versions of a standard form for use with domestic sub-contracts (one for use where a design obligation is to be sub-contracted, and one for use where it is not) and although there is no requirement under SBC05 that the main contractor should use this form, there are restrictions on the terms that may be agreed. These are set out in clause 3·9·2 of SBC05, which requires, for example, that particular Conditions relating to ownership of unfixed goods and materials, and the right to interest on unpaid amounts properly due to the sub-contractor, are included in all domestic sub-contracts. The sub-contract should also, of course, comply with the requirements of the Housing Grants, Construction and Regeneration Act 1996 (HGCRA 1996).

Use of documents

Interpretation, definitions

3.27   Section 1 of the form sets out some rules governing the interpretation of the Conditions. Clause 1·1 is a schedule of definitions of terms that are used throughout the contract.

Some further and more detailed definitions are embodied in the text of clauses, for example 'All Risks Insurance' and 'Joint Names Policy' are defined in clause 6·8. Clause 1·4, which is substantially new to the 2005 edition of the form, defines what is meant by any reference to a 'person', or to a statute, and also contains a gender bias clause. Section 1 also includes items first introduced by Amendment 18 to JCT80. Those regarding notices and periods of time were required by the Housing Grants, Construction and Regeneration Act 1996 (HGCRA 1996) and re-state its requirements relating to the serving of notices and the calculation of periods of days (cl 1·5 and 1·7). Clause 1·8 allows the parties to agree that certain communications, to be identified in the Contract Particulars, may be made electronically. The parties may also agree an exact format for the electronic communications. Otherwise, all communications are to be in writing.

Priority of contract documents

3.28   Clause 1·3 states 'the Agreement and these Conditions are to be read as a whole but nothing contained in the Contract Bills or the CDP Documents shall override or modify the Agreement or these Conditions'. If this clause was not included, the position under common law would be the reverse; in other words, anything that had been specifically agreed and included in a document would normally over-ride any standard provisions in a printed form.

3.29   If the parties wish to agree to any special terms that differ in any way from the printed Conditions, then the amendments will need to be made to the actual form, which usually involves the insertion of one or more additional articles. If necessary, due to lack of space, these amendments could refer to the special terms, which could be appended to the form or included in the bills of quantities. Amending standard forms is unwise without expert advice as the consequential effects are difficult to predict. Deleting clause 1·3 could be particularly unwise as it may have unintended effects on other parts of the contract. (If significant changes are needed, consider use of another form, perhaps the Major Project Form which embraces individual tailoring and has no equivalent clause.)

Inconsistencies, errors or omissions

3.30   The contractor is under an obligation to point out any discrepancy or divergence within or between the contract documents, including the CDP documents, and/or any further instructions, documents or drawings issued by the contract administrator (cl 2·15). The obligation appears to be limited to those discrepancies that the contractor has discovered. There is no obligation for the contractor to search for discrepancies, although the general obligation to use reasonable skill and care would suggest some degree of observance could be expected. Any notice should be issued immediately upon discovery and should include 'appropriate details' of the error or discrepancy. If the contractor fails to point out any discrepancies that it notices, or should have noticed, and work has to be re-done as a result, then the contractor may lose any right to extra payment, extension of time, and loss and expense. The contract administrator's obligation to issue an instruction under clause 2·15 appears to be limited to instances where the contractor has found a discrepancy. However, the contract administrator's general obligation to provide necessary information would extend to correcting any errors and discrepancies.

Errors in the contract bills

3.31   The contract requires that any error in the contract bills shall be corrected (cl 2·14·1). It does not say by whom but this would be the responsibility of the employer and normally carried out by the quantity surveyor. The correction is treated as if it were a variation required by a contract administrator's instruction (cl 2·14·3), and though there is no express requirement to issue such instruction, the correction should be confirmed in writing. The wording is different in the Without Quantities version, where clause 2·14 requires the contract administrator to issue instructions in regard to inconsistencies, errors or omissions.

The contract bills and the drawings

3.32   The contract does not specifically deal with the situation where there is a divergence between the information shown in the contract bills and that set out on the contract drawings. However, as clause 4·1 states that 'the quantity and quality of the work included in the Contract Sum shall be deemed to be that set out in the Contract Bills', in the case of conflict the bills will normally take precedence. If some other result is preferred then the contract administrator will need to issue an instruction, which will constitute a variation.

Employer's Requirements and Contractor's Proposals

3.33   Where there is an error in the Contractor's Proposals or the CDP analysis, this is corrected, but is not to result in any addition to the contract sum (cl 2·14·4). The contractor must inform the contract administrator of its proposed amendment to deal with any discrepancy between the CDP documents other than the Employer's Requirements. The contractor is obliged to accept the contract administrator's decision and comply at no cost to the employer (cl 2·16·1). If the contract administrator failed to reach a decision within a reasonable time, this could be grounds for an extension of time and loss and/or expense.

3.34   Where there is a discrepancy within the Employer's Requirements, or a discrepancy between the Requirements and any variation, the contract states that if the Contractor's Proposals deal with the discrepancy then they will prevail (cl 2·16·2). The discrepancy between the Requirements and any variation refers to inadvertent problems resulting from the effect of a variation, rather than intended alterations to the particular part of the Requirements at which the variation was aimed. If the employer decides it does not like the solution in the Contractor's Proposals, and would prefer some other solution, this would have to be instructed as a variation.

3.35   If the Proposals do not deal with the discrepancy, the contractor is required to inform the employer of its proposed amendment for dealing with it, and the contract administrator must either agree or decide on alternative measures and in either case notify the contractor in writing (cl 2·16·2). The acceptance or notification is to be 'treated as a variation', which would result in it being valued under clause 5·2, and in it constituting grounds for an extension of time under clause 2·29·1, for loss and/or expense under clause 4·24·1, and for termination under 8·9·2·1, in the unlikely event that it causes a suspension (see Chapter 6). If there was undue delay by the contract adminstrator in reaching a decision, then this would also be grounds for a claim.

3.36   The contract does not deal with the situation where there is a divergence between the Employer's Requirements and the Contractor's Proposals. The tenth recital states 'the Employer has examined the Contractor's Proposals and the CDP Analysis and, subject to the Conditions and, where applicable, the Supplementary Provisions hereinafter contained, is satisfied that they appear to meet the Employer's requirements'.

3.37   This recital is somewhat problematic from the point of view of the employer. Although not a condition of the contract it nevertheless, in the absence of any contrary provision, appears to give precedence to the Contractor's Proposals and the CDP Analysis. The notes to the WCD98 third recital in CD1/B state that the intention of this recital is that it should be without prejudice to the contractor's liability in respect of design, and that, for example, if the Employer's Requirements included a performance specification for a heating system, and the employer subsequently accepted the contractor's design proposals for the system, the employer would not be precluded from alleging breach of contract. This may have been the intention of the drafters but the position is far from clear.

3.38   If there was a discrepancy between the CDP documents, a court will endeavor to determine, from an objective standpoint, what were the true intentions of the parties, and might well decide that the employer should be deemed to have accepted the version set out in the Proposals at least in so far as the divergence would have been revealed by a reasonably thorough examination. As the employer may prefer its own Requirements to take precedence, this recital is frequently altered to that effect in practice.

Divergences from statutory requirements

3.39   The contractor and the contract administrator are both required to notify each other of any discrepancy or divergence between any of the clause 2·15 documents, or any instruction requiring a variation, and any statutory requirement as defined under clause 1·1 (cl 2·17·1). Where the discrepancy relates to the CDP documents, the contractor must inform the contract administrator of its proposed amendment to deal with the discrepancy. In all cases the contract administrator must issue instructions to deal with the problem. Where the diveregence relates to the CDP documents, the contract states that the contractor must comply at no extra cost to the employer, unless it results from a change in statutory requirements since the Base Date. In all other cases the instruction is treated as a variation (cl 2·17·2). The effect of this clause is that the costs will be borne by the contractor in situations where the divergence is between the Employer's Requirements and statute, as well as between the Contractor's Proposals and statute.

Custody and control of documents

3.40   The contract drawings and contract bills remain in the custody of the employer, and must be available for inspection at all reasonable times (cl 2·8·1). The contract administrator should retain a copy for reference throughout the life of the contract. The contractor must be provided with one certified copy and two further copies (cl 2·8·2).

3.41   The documents provided must not be used for any purpose other than the Works, and the details of the rates or prices are not to be divulged (cl 2·8·4). The contractor must keep on site at all reasonable times one copy of all the contract documents, the unpriced bills of quantities, the master programme, and further schedules and information issued by the contract administrator (cl 2·8·3). The clause does not include a reference to the contractor's Design Documents prepared during the course of the Works under clause 2·9·2, and it may be wise to clarify whether or not this is required in the tender documents. At practical completion the contractor must provide copies of drawings and information relating to the Contractor's Designed Portion as stipulated in the contract documents, or as the employer may reasonably require (clause 2·40).

Assignment and third party rights

Assignment

3.42   The right of a subsequent purchaser to bring an action against the builder of their property, with whom they have had no contractual relationship, could be of considerable value. The employer in a construction contract might therefore wish to assign this right to such other person who may acquire an interest in the property.

3.43   A contractual right can be regarded as a personal right of property, and in property law it is classified as 'chose in action'. Choses in action can be assigned under the Law of Property Act 1925, provided the requirements of section 136 of the Act are followed. It is important to note that it is only contractual rights that can be assigned, termed 'the benefit' of a contract, and not obligations. So if, for example, A enters into a contract with B whereby A agrees to carry out some building work, and B agrees to pay A £100 for the work, A can assign the right to claim the £100 to C but not the obligation to carry out the work. The right to pursue a debt or claim is assignable to C without B's consent, provided B is notified as required by section 136. The obligation to carry out the work, however, could only be transferred to C with the agreement of all three parties (often termed 'novation').

3.44   SBC05 contains express provisions which limit the scope for assigning contractual rights. Clause 7·1 states that neither the employer nor the contractor may 'assign this Contract or any rights thereunder' without the written consent of the other. Assignment without consent of the other party is grounds for termination (cl 8·4·1·4 and 8·9·1·3). There is one exception, however, to the prohibition on assignment: if clause 7·2 is stated to apply in the Contract Particulars then the employer may assign some limited rights to a party to whom it has transferred a freehold or leasehold interest in the premises comprising the Works. Among other limitations the rights can only be assigned after practical completion. The clause does not provide a general right to assign the benefit, but the right to bring proceedings in the name of the employer to enforce terms of the contract made for the benefit of the employer. It is thought that this would limit the assignee to claiming at most losses suffered by the employer as a result of any breach by the contractor, and would not extend to further losses suffered by itself.

Third party rights/warranties

3.45   SBC05 offers two options for the granting of rights to bring a claim to persons who are not a party to the contract, either through the use of the 'third party rights' provisions included in the form, or through the use of separately published standard form warranties.

3.46   The 'third party rights' provisions make use of the facility introduced relatively recently by the Contracts (Rights of Third Parties) Act 1999. Until this Act came into force, it was a rule of English law that only the two parties to a contract had the right to bring an action to enforce its terms (termed 'privity of contract'). However, it is often the case in construction projects that other parties may wish to be in a position to be able to take action, should one or other of the parties default on their obligations. A future owner of the property may, for example, wish to be able to claim against the contractor should it later transpire that the project was not built according to the contract. Under the rule of privity, the future owner would be a third party, and would not be able to bring a claim. In response to this, 'collateral warranties' were developed which allowed for third parties to pursue claims for breaches of a contract. Examples of such warranties would be between contractor and owner, contractor and funder, and also between consultants and owners/funders.

3.47   The Contracts (Rights of Third Parties) Act has changed the fundamental rules of law relating to privity, in that it entitles third parties to enforce a right under a contract, where the term in question was to provide a benefit to that third party. The third party could be specifically named, or could be of an identified class of people. The effect of this Act is therefore to open the door to the possibility of claims being brought by a range of persons, in some cases persons that the parties to the contract may never have considered.

3.48   The Act, however, allows for parties to agree that their contract will not be subject to its provisions, and many standard forms adopt this course in order to limit the parties' liability. SBC05 takes this approach and under clause 1·6 states:

> Other than such rights of any Purchasers, Tenants and/or Funders as take effect pursuant to clauses 7A and/or 7B, nothing in the Contract confers or is intended to confer any right to enforce any of its terms on any person who is not a party to it.

3.49   The contract therefore by this clause 'contracts out' of any effects of the Act. (In the light of the above, it is important to note that the effects of deleting or amending this clause would be significant.) It then replaces it with a set of terms which define precisely which third parties will have rights with respect to the contract (Part 2 of the Contract Particulars), and what those rights will be (Schedule 5).

3.50   Schedule 5 sets out 'Third Party Rights for Purchasers or Tenants' (Part 1) and 'Third Party Rights for a Funder' (Part 2). The contractor warrants (in relation to the tenant) that it has carried out the Works in accordance with the contract (with effect from practical completion), and (in relation to the funder) that it has complied with and will continue to comply with the contract. This allows both the purchaser/tenant and the funder to bring an action in respect of breaches of contract by the contractor.

3.51   There are some things to note about this system. In the case of purchasers and tenants, the contractor's liability extends to the reasonable costs of repair, renewal or reinstatement, but does not include other losses unless so stated in the Contract Particulars (Schedule 5: Part 1, cl 1·1), in which case the liability will be limited to a stated maximum amount. The contractor's liability is also limited by a net contribution clause (Schedule 5: Part 1, cl 1·3). Under the arrangement the contractor is entitled to rely on any term in the contract should any action be brought against it by a third party (Schedule 5: Part 1, cl 1·4). Where there is a Contractor's Designed Portion, the contractor is required to provide evidence of its professional indemnity insurance to any person possessing rights under the Third Party Rights Schedule (Schedule 5: Part 1, cl 1·8). The rights may be assigned by the purchaser or tenant without the contractor's consent to another person, and by that person to a further person, but beyond this no further assignment is permitted (Schedule 5: Part 1, cl 1·7).

3.52   In the case of the funder, except for the inclusion of a net contribution clause (Schedule 5: Part 2, cl 1·1), no limit is placed upon the extent of the contractor's liability. As above, the contractor is entitled to rely on any term in the contract should any action be brought by the funder (Schedule 5: Part 1, cl 1·2), and the rights may be assigned by the funder without the contractor's consent to another person, and by that person to a further person, but beyond this no further assignment is permitted (Schedule 5: Part 2, cl 1·10). The Schedule also sets out various 'stepping in' rights which may be exercised by the funder in the event that it terminates its Finance Agreement with the employer.

3.53   Under the alternative system of 'collateral warranties' the contractor has actually to enter into a warranty separately with each beneficiary. The beneficiaries are identified in Part 2 of the Contract Conditions, and the warranties are identified in clause 7C and 7D as the JCT standard forms of warranty to purchaser/tenant and funder (CWa/P&T, and CWa/F). At the time of writing these warranties are not yet available, but it is understood from the JCT Guide that they comprise identical terms to the third party rights set out in Schedule 5.

Procedure with respect to third party rights and warranties

3.54   Where third party rights are to be used, the relevant sections of the Contract Particulars must be completed carefully (Part 2: A–D, and clauses 7A and 7B). In the case of warranties from the contractor, Part 2: A–D, and clauses 7C and 7D, must be completed. It is important to identify the funder/purchaser/tenant because if none is identified the rights/warranties shall not be required. It is not necessary, however, to identify a specific organisation; the description could simply be of a class of persons, e.g. 'all first purchasers' or 'the lead bank providing finance for the project'.

3.55   The third party rights take effect from the date of receipt by the contractor of the employer's notice to that effect, in the case of a purchaser or tenant it must state their name and their interest in the works, and in the case of a funder simply identifying the party concerned. Where collateral warranties are required, the contractor is required to execute the stipulated warranties within 14 days of the equivalent notice from the employer.

3.56   From the point of view of the purchasers, tenants or funders, they will not be aware of the existence of the third party rights unless the employer lets them have a copy of the relevant part of the contract. In some cases the third party may prefer to have a separate collateral warranty direct with the contractor, and it would be sensible of the employer to establish whether this may be a possibility before executing the main contract. After the contract is executed, this could only be arranged with the consent of the contractor.

3.57   With respect to warranties from sub-contractors, Part 2: E must be completed (clauses 7E and 7F). The requirement for obtaining warranties (clause 7F) states that:

> Where Part 2 of the Contract Particulars provides for the giving by any sub-contractor of a Collateral Warranty to the Employer, the Contractor shall within 21 days from receipt of the Employer's notice identifying the relevant sub-contractor comply with the requirements as set out in the Contract Documents as to obtaining such warranties subject to any amendments proposed by any such sub-contractor and approved by the Contractor and the Employer, such approval not to be unreasonably delayed or withheld.

3.58   It should be noted that this is not, of itself, a requirement for the contractor to obtain the warranties, nor is it even (as with GCWks1) a requirement to use reasonable endeavours to obtain the warranties – it is simply a requirement to comply with the contract Conditions. The contractor is, however, required to include provisions as necessary in sub-contracts in respect of the execution of required warranties (cl 3·9·2·40). No standard form for a warranty is referred to (nothing is included under Schedule 5 to cover this). The JCT do, however, intend to publish a standard form of sub-contract warranty (SBWa/E) to cover this situation.

4 Contractor's obligations

4.1 The contractor's paramount obligation is to 'carry out and complete the Works'. This is stated in article 1 and reinforced in clauses 2·1 and 3·6, the latter being a clear statement that the contractor is held wholly responsible for achieving this, irrespective of whether the contract administrator or clerk of works visits or is present on the site.

The Works

4.2 The Works that the contractor undertakes to carry out will be as briefly described in the first recital of SBC05, and as shown or described in the contract documents. It is therefore important to check that the entry in the first recital clearly identifies the nature and scope of the proposed work, and that descriptions of the Works given elsewhere are clear and adequate. Under the seventh recital the employer may stipulate that the Works include the design and construction of an identified part or parts of the project, termed 'the Contractor's Designed Portion'. The term 'Works' is consequently defined under clause 1·1 as 'including, where applicable, the CDP Works'.

4.3 Note also that as defined in clause 1·1, the Works will also include any changes subsequently brought about by a contract administrator's instruction, which might also introduce additional drawings or other information. These might not be 'contract documents', but they nevertheless have an important status and the contractor is obliged to carry out any additional work which they show.

Contractor's design obligation

4.4 The contractor's design obligation is set out under clause 2·2, which states 'where the Works include a Contractor's Designed Portion the contractor shall … in accordance with the Contract Drawings and the Contract Bills (to the extent that they are relevant) complete the design for the Contractor's Designed Portion'.

4.5 The design requirements will have been set out in the Employer's Requirements and sent out with the tender documents (eighth recital). The contractor will have submitted a proposal containing a design solution with its tender (the Contractor's Proposals, ninth recital) although, depending on the information requested, this may not be fully detailed. Some of the design may therefore remain to be finalised after the contract is entered into. Somewhat surprisingly, clause 2·2 does not state that the design should be completed in accordance with the Employer's Requirements. Although it is likely that this would be implied, it may be wise to clarify this in the tender documents, possibly by including the Requirements as a section of the bills.

4.6 Clause 2·13·2 makes it clear that the contractor is not responsible for the contents of the Employer's Requirements, or for verifying the adequacy of any design contained within them. This clause is included to prevent such an obligation being implied, as it was in the case of *Henry Boot* v *Co-Op Insurance Society*. Although it is not entirely clear, it is unlikely to prevent the implication of a 'duty to warn' regarding any other aspects of the consultant team's design, for example where the design is varied through an instruction (for an example of this see the earlier case of *Plant Construction v Clive Adams*).

4.7   As discussed above, the tenth recital states 'the Employer has examined the Contractor's Proposals and, subject to the Conditions, is satisfied that they appear to meet the Employer's Requirements', which implies that in so far as the design has been finalised at the time of acceptance of tender, then the employer has accepted the solution. It might be possible to argue that the employer could not be held to have accepted defects in the design, which a reasonable inspection would not have revealed. An example might be the design of a roof truss, where without a detailed double checking of calculations it would not be possible to ascertain whether the truss would be structurally sound. The contractor would therefore remain responsible for achieving this whatever the Proposals showed. Nevertheless, the recital is problematic from the point of view of the employer and is sometimes deleted.

Plant Construction v Clive Adams Associates and JMH Construction Services [2000] BLR 158 (CA)

Ford motor company engaged Plant on a JCT WCD contract to design and construct two pits for engine mount rigs at Ford's research and engineering centre in Essex. Part of the work included underpinning an existing column, and in the course of the work temporary support was required to the column and the floor above. JMH were sub-contracted to carry out this concrete work. Ford's own engineer gave instructions regarding the temporary supports, which comprised four Acrow props. JMH and Plant's engineers, Clive Adams, felt the props to be inadequate and discussed this on site. The support was installed as instructed and failed, so that a large part of a concrete floor slab collapsed. Plant settled with Ford, and brought a claim against JMH and Clive Adams (who settled). The court found that it was part of the duties of the sub-contractor to warn of any aspect of the design it knew to be unsafe. It reserved its opinion on whether the duty would extend to unsafe aspects it ought to have known, or design errors that were not unsafe.

Co-operative Insurance Society v Henry Boot Scotland and others [2002] CILL 1932

The Society engaged the contractor Henry Boot on an amended version of JCT80 incorporating the Design Portion Supplement, where the relevant terms are virtually identical to that of WCD98. During construction, problems arose where soil and water flooded into a basement excavation. An engineer had originally been employed by the Society to prepare a concept design for the structure, and Henry Boot had developed the design and prepared working drawings. The Society brought claims against Henry Boot and the engineers. Henry Boot argued that their liability was limited to the preparation of the working drawings. The judge, however, took the view that completing the design of the contiguous bored pile walls included examining the design at the point that it was taken over, assessing the assumptions on which it was based and forming a view as to whether they were appropriate.

Ad-hoc approaches

4.8   Contract administrators sometimes attempt to place a design obligation on the contractor through clauses in the bills or by a reference in other documents, or through the inclusion of a performance specification in a description of the Works, other than the Contractor's

Figure 6 Watchpoints: Contractor's Design

- The contractor's liability for providing the CDP work is limited to the use of reasonable skill and care (cl 2·19·1)
- Level of PII insurance must be stated in the Contract Particulars, otherwise none will be required (cl 6·11·1)
- It is unclear what level of design responsibility the contractor will have for any design not stated to be included in the Contractor's Designed Portion, or whether the contractor is required to insure for this
- In cases where the Contractor's Proposals are found not to comply with the Employer's requirements, it is unclear which takes precedence (tenth recital, cl 2·16)
- Integration of the design work remains the responsibility of the contract (cl 2·2·2)
- The tender documents should state the exact scope and format of the information to be included in the Contractor's Proposals
- The contractor is obliged to submit further 'reasonably necessary' information 'as and when necessary' (cl 2·9·2 and 2·9·3). It is suggested that the exact information required and dates for submission are set out in the contract documents
- Information required to be submitted at practical completion should be set out in the contract documents (cl 2·40)
- The Employer may make changes to the Employer's Requirements (which may result in changes to the contract sum and the completion date), but otherwise there is no power to order changes to the Contractor's Proposals provided they comply with the Requirements (Schedule 1).

Designed Portion. It would be unwise to try to assign a design role in this way as the outcome cannot be predicted with certainty. The new wording of clause 2·3 gives some support to the argument that the contractor would be responsible for providing something 'appropriate to the works' (see paragraph 4.18 below) and there have been cases where a court has found ad-hoc methods have placed a design obligation on the contractor. However these did not involve a form which made provision for a 'Contractor's Designed Portion'. The outcome with SBC05 will no doubt depend on the particular circumstances of the case.

The level of design liability

4.9 Standard forms of contract and appointment will often set out specific provisions regarding design liability, but these have to be understood in the legal context in which they operate. A key point is whether any design liability incurred is a 'fitness for purpose' or 'reasonable skill and care' level of liability.

4.10 The Sale of Goods Act 1979 implies terms into all contracts for the sale of goods that the goods sold will be of satisfactory quality. The Unfair Contract Terms Act 1977 stipulates that this requirement cannot be excluded in any contract with a consumer, and can only be excluded in other contracts in so far as it would be reasonable to do so. If parties have included terms which purport to exclude this liability, the terms will be void. Similarly, if the use to which the goods are to be put is made clear to the seller, the seller must supply goods suitable for that use unless it is clear that the buyer is not relying on the seller's skill and judgement. So if, for example, a DIY enthusiast asks a builder's merchant for paint suitable for use on a bathroom ceiling, the merchant must supply suitable paint, regardless of what is written in the contract of sale. If, however, the buyer specifies the exact type of paint, the seller would no longer be liable as the buyer is not relying on the seller's advice.

4.11   Contracts for construction work are usually for 'work and materials' (as opposed to supply-only or install-only) and as such fall under the Supply of Goods and Services Act 1982. This implies similar terms to those described above in relation to any goods supplied under such a contract. Therefore a contractor would normally be liable for providing materials fit for their intended purposes. If, however, an employer or consultant specifies particular materials, the contractor would be relieved of this liability.

4.12   The obligation to supply goods or materials fit for their intended purpose would extend to a product or structure which a contractor had agreed to design and construct (*Viking Grain Storage Ltd v T. H. White and another*). In all cases the liability of the contractor will be strict; in other words, the contractor will be liable if the goods, element or structure is not fit for its intended use, irrespective of whether the contractor has exercised a reasonable level of skill and care in carrying out the design. This is a more onerous level of liability than that assumed by someone undertaking design services only, where they would normally be required to demonstrate that they had exercised the skill and care of a competent member of their profession. To put it the other way round, if an employer can prove that a building designed and constructed by a contractor is defective, then this will normally be sufficient to prove that there has been a breach of contract, whereas in the case of a design professional the employer would also have to prove that the professional had been negligent.

Viking Grain Storage Ltd v T. H. White and another (1985) 33 BLR 103

Viking Grain entered into a contract with White to design and erect a grain drying and storage installation to handle 10,000 tons of grain. After it was complete, Viking commenced proceedings against the contractor claiming that, because of defects, the grain store was unfit for its intended use. The contractor in its defence claimed that there was no implied warranty in the contract, that the finished product would be fit for purpose, and that the contractor's obligation was limited to the use of reasonable skill and care in carrying out the design. The judge decided that Viking had been relying on the contractor and because of this there was an implied warranty that not only the materials supplied but also the whole installation should be fit for the required purpose. There could be no differentiation between reliance placed with regard to the quality of the materials and to the design.

4.13   Under clause 2·19·1 of SBC05, the contractor's liability for the Contractor's Designed Portion is limited to using the skill and care of 'an appropriately qualified and competent professional designer'. In effect, this means that in order to prove a breach the employer would need to prove that the contractor had been negligent. If, for example, the contractor is required to design a heating system to heat the rooms to a certain temperature, and when installed it fails to do so, this fact alone would not be enough to prove that there had been a breach of contract. The employer would need to prove that the contractor had failed to use the skill and care expected of a professional person.

4.14   It should be noted that where the contractor is carrying out work in connection with a dwelling, including design work, this would be subject to the Defective Premises Act 1972. This obligation is acknowledged in clause 2·19·2. The Act states that 'a person taking on work in connection with the provision of a dwelling ... owes a duty ... to see that the work which he takes on is done in a workmanlike or, as the case may be, professional manner, with proper materials and so that as regards that work the dwelling will be fit for habitation when completed' (s.1(1)). This appears to be a strict liability, and is owed to anyone acquiring an interest in the dwelling. So although the contractor's liability to the employer under clause 2·19·1 may be limited to the use of due skill and care, it should remember that it may acquire a higher level of liability to third parties under statute. In addition, although the contractor's liability is limited to the amount stated in the contract particulars, the limitation does not apply to work in connection with a dwelling (cl 2·19·3).

Materials, goods and workmanship

4.15   Work must be carried out in a proper and workmanlike manner and in accordance with the Health and Safety Plan (cl 2·1). Any failure in these respects is regarded as serious, and the contract administrator has the power to intervene by issuing an instruction if necessary (cl 3·19).

4.16   Clause 2·3·1 states that all materials and goods shall be of the standard described in the contract bills (or in the specification in the Without Quantities version). The obligation is qualified by the phrase 'so far as procurable'. It would be an implied duty that the contractor should notify the contract administrator before substituting any materials or goods instruction, even where those specified are unobtainable. The substitution would result in a variation that should be covered by a contract administrator's instruction. With respect to the Contractor's Designed Portion, the standard of materials and goods should be that in the Employer's Requirements or, if none is given, then that shown in the Contractor's Proposals or in further drawings supplied by the contractor.

4.17   Standards of workmanship should be as specified in the contract bills (or specification, cl 2·3·2). In respect of the Contractor's Designed Portion, the standard of workmanship is that set out in the Employer's Requirements or, if none is set out, is that set out in the Contractor's Proposals.

4.18   Clause 2·3·3 states:

> Where and to the extent that the approval of the quality of materials or goods or of the standards of workmanship is a matter for the opinion of the Architect/ Contract Administrator, such quality and standards shall be to his reasonable satisfaction. To the extent that the quality of materials and goods or standards of workmanship are neither described … nor stated to be a matter for such opinion or satisfaction, they shall in the case of the Contractor's Designed Portion be of a standard appropriate to it and shall in any other case be a standard appropriate to the Works.

4.19   This reflects the duty that would normally be implied by law, in other words where the description of the standard required for any goods, materials and workmanship is (deliberately or inadvertently) incomplete, the contractor is required to provide something 'fit for purpose' (see paragraph 4.10). This appears to be a strict obligation (see above), rather than an obligation to use reasonable skill and care.

4.20   The phrase 'where and to the extent that approval of the quality … is a matter for the opinion of the contract administrator, such quality and standards shall be to the reasonable satisfaction of the contract administrator' (cl 2·3·3) does not authorise the contract administrator to alter the standard specified at will, but means that where a correct construction of the contract documents leaves a matter regarding quality to the discretion of the contract administrator, the contractor only fulfils its obligations if the contract administrator is satisfied. Any expression of dissatisfaction by the contract administrator must be made within a reasonable period of the carrying out of the work, and the contract administrator must state reasons for the dissatisfaction (cl 3·20). It should be noted that the Final Certificate is conclusive evidence that where the contract documents have expressly stated that the quality is to be to the approval of the contract administrator, then the contract administrator is so satisfied (cl 1·10·1). This would have the effect of preventing the employer bringing a claim regarding those items of work. The contract administrator should generally avoid using phrases such as 'to approval' or 'to the contract administrator's satisfaction' in the contract documents.

4.21   If the phrase 'or otherwise approved' is used in a specification or bill of quantities this does not mean that the contract administrator must be prepared to consider alternatives put forward by the contractor, nor that the contract administrator must give any reasons for rejecting alternatives (*Leedsford* v *City of Bradford*). It merely gives the contract administrator the right to do so. A substitution would always constitute a variation whether or not this phrase is present in the specification.

Leedsford Ltd v The Lord Mayor, Alderman and Citizens of the City of Bradford (1956) 24 BLR 45 (CA)

In a contract for the provision of a new infant school the contract bills stated 'Artificial Stone… The following to be obtained from the Empire Stone Company Limited, 326 Deansgate, or other approved firm…'. During the course of the contract the contractor obtained quotes from other companies and sent them to the contract administrator for approval. The contract administrator, however, insisted that Empire Stone was used and as Empire Stone was considerably more expensive the contractor brought a claim for damages for breach of contract. The court dismissed the claim stating:

> The builder agrees to supply artificial stone. The stone has to be Empire Stone unless the parties agree some other stone, and no other stone can be substituted except by mutual agreement. The builder fulfils his contract if he provides Empire Stone, whether the Bradford Corporation want it or not; and the Corporation Contract administrator can say that he will approve of no other stone except the Empire Stone. (Hodson LJ at page 58).

Obligations in respect of quality of sub-contracted work

4.22   With increasing specialisation in the construction industry it is almost universal practice for much of the work on building projects to be sub-contracted to a large number of other firms. This arrangement benefits the employer by enabling it to take advantage of a wider range of specialisms than would normally be available within one contracting organisation. The employer will wish, nevertheless, to be able to hold the contractor responsible to a degree for any non-performance of the sub-contractors.

4.23   There are two methods of sub-contracting allowed for under SBC05:

- sub-letting to a domestic sub-contractor selected by the main contractor, but with the written consent of the contract administrator (cl 3·7);

- sub-letting to a domestic sub-contractor selected from a list of three names under the procedure set out in clause 3·8.

4.24   Both of these methods are discussed in detail in Chapter 6. With respect to quality, in both cases the contract makes it clear that the contractor remains entirely responsible for the performance of any sub-contractor (cl 3·7·1). This responsibility would extend to any Contractor's Designed Portion work that was sub-contracted. As discussed at paragraph 4.8, if any ad-hoc method is adopted, such as attempting to 'nominate' a sub-contractor in the Contract Bills or specification, then responsibility for quality and fitness will depend on the particular facts of each case. It should be noted, however, that attempts to hold the contractor liable based on early editions of the standard form were not always successful (see for example *Gloucester County Council v Richardson*).

Gloucester County Council v Richardson [1968] 1 AC 480 (CA)

Under a building contract on JCT37 (1957 edition) concrete columns, which the contractor was to supply under it, were to be ordered from suppliers nominated by the employers. A PC sum had been stated in the Bills of Quantities, which named a particular supplier. On accepting the tender, the architect nominated a different supplier, who it transpired would only contract on particular terms which excluded all liability for defects or their damaging consequences save only an obligation to replace. The columns had defects which were undetectable when they were supplied, but which appeared after some columns had been incorporated in the building under construction. Once the defects were discovered the contractor was told to stop work on the columns, which eventually led to the contractor determining the contract.

In an action by the employers against the contractor, the employers argued that the contractor had repudiated the building contract, and that the defects in the concrete columns were a breach of its implied warranty of fitness. The court found that the contractor was not liable for the defects, commenting thay it felt it would be unfair. Russel LJ pointed out that that 'if the employers wished to impose liability on the contractor for defects in materials supplied by nominated suppliers they could do so in plain terms, such as are contained in clause 31 of the General Conditions of Government Contracts (October 1959)'.

Compliance with statute

4.25   The contractor is under a statutory duty to comply with all legislation that is relevant to the carrying out of the work, for example in respect of goods and services, building and construction regulations, and health and safety. The duty is absolute and there is no possibility of contracting out of any of the resulting obligations.

4.26   SBC05 imposes a contractual duty in addition to the statutory duty. Under clause 2·1 the contractor is required to carry out and complete the works 'in compliance with the Health and Safety Plan and the statutory requirements'. The contractor is required to pay all fees and charges but unless these have already been included in the contract sum or are covered by a provisional sum, the amounts will be added to the contract figure (cl 2·21).

4.27   If the contractor finds any divergence between what the contract requires and statutory requirements, then the contract administrator must be given immediate written notice

Figure 7 Key duties of the contractor

2·1	Carry out and complete the works
2·2·1	Complete the design of the CDP
2·2·2	Comply with CA's directions regarding integration of the CDP
2·2·3	Comply with regulation 13 of the CDM Regulations
2·3·1	Provide materials and goods of standards described
2·3·2	Provide workmanship and goods of standards described
2·3·3	Provide materials goods and workmanship to satisfaction of CA
2·3·4	Provide CA with vouchers
2·3·5	Take all reasonable steps to encourage contractor's persons to be CSCS registered cardholders
2·4	Retain possession of the site
2·6·1	Notify the employer of additional prelims
2·7·1	Permit execution of work by employer's persons
2·8·3	Keep copy of documents on site
2·9·1·2	Provide copies of programme and amended programme
2·9·2	Provide copies of Design Documents, and levels and setting out
2·15	Notify CA of discrepancies
2·16·1	Notify CA of discrepancies in relation to CDP documents
2·16·2	Inform the CA of amendment to deal with discrepancies
2·17·1	Notify the CA of any discrepancy between documents and statutory requirements
2·18·1	Execute work in relation to emergency compliance with statutory requirements
2·18·2	Inform the CA of 2·18·1 work
2·21	Pay and indemnify the employer against liability for statutory fees
2·22	Indemnify the employer against liability for royalties and patent rights
2·27·1	Notify CA of delays and their causes
2·27·2	Notify CA of further particulars of delays
2·27·3	Notify CA af any material change in the estimated delay
2·38	Make good defects notified by CA
2·41	Supply as-built drawings

Figure 7 Continued

3·1	Secure a right of access for CA to sub-contractors' workshops
3·2	Ensure a competent person-in-charge is on site at all times
3·8·3	Add persons to the list of sub-contractors
3·10	Comply with instructions issues by architect
3·10·3	Notify the CA of the injurious effect of an instruction
3·12·1	Confirm oral instruction
3·22·1	Use best endeavours not to disturb object of antiquity
3·22·2	Take all steps necessary to preserve object
3·22·3	Inform architect of discovery of object
3·25	Comply with CDM Regulations
3·25·2	Provide employer with Health and Safety Plan and notify employer of any changes
3·25·3	Provide planning supervisor with information for Health and Safety File
4·5·1	Provide CA with documents for final adjustment of the contract sum
4·8	Provide a bond in relation to any advance payment
4·17·4	Provide a bond in relation to off-site materials
4·17·5	Provide a bond in relation to off-site materials
4·19·2	Provide a bond in lieu of retention
4·19·4	Arrange for increase in amount stated in bond
5·3·1	Provide a schedule 2 quotation
6·1	Indemnify the employer against losses, etc., due to personal injury or death of any person
6·2	Indemnify the employer against losses, etc., due to damage to property other than the works
6·4·1	Take out and maintain insurance in respect of 6·1 and 6·2 liability
6·4·2	Send evidence of insurance to the CA
6·5·1	Take out and maintain insurance in respect of damage to property caused (without negligence) by carrying out of works
6·5·2	Send 6·5·2 policy to CA
6·9·1	Ensure joint names policy provides for recognition of, or waives any right of subrogation against, any sub-contractor

Figure 7 Continued

6·10·1	Notify the employer if notified that terrorism cover has ceased
6·10·4·1	Restore work damaged by terrorism
6·11	Take out and maintain professional indemnity insurance, and provide evidence that insurance has been effected
6·12	Give notice to employer if 6·11 insurance ceases to be available
6·14	Comply with the Joint Fire Code, and ensure all contractor's persons comply
7C	Enter into a warranty with a purchaser or tenant
7D	Enter into a warranty with a funder
7E	Comply with requirements set out in contract particulars as to obtaining sub-contractor warranties with purchasers and tenants/funders
7F	Comply with requirements set out in contract particulars as to obtaining sub-contractor warranties with the employer
8·5·2	Give the employer notice if it makes any proposal, etc., in relation to insolvency
8·7·2·1	Remove temporary buildings etc. from the site
8·7·2·2	Provide employer with two copies of all design documents
8·7·2·3	Assign to the employer the benefit of any agreements to supply materials etc. or execute work
8·7·5	Pay the employer any balance due
8·12·2·1	Remove temporary buildings etc. from the site
8·12·2·2	Provide employer with two copies of all design documents
8·12·3	Prepare an account or provide employer with documents to prepare an account
Schedule 1	
1	Prepare and submit two copies of each design document to the CA
5·1	Carry out CDP works
5·2	Incorporate comments and carry out CDP works
Schedule 2	
1·2	Submit schedule 2 quotation
7	Confirm any agreement to adjust time period for acceptance of quotation
Schedule 3	
A·1	Take out and maintain a joint names policy for 'All Risks Insurance'

Figure 7 Continued

A·2	Send policy to CA, together with each premium receipt and any policy endorsements
A·4·1	Give notice to CA in the event of occurrence of loss or damage covered
A·4·3	Restore the damaged work
A·4·4	Authorise the insurers to pay all monies to the employer
B·3·1	Give notice to CA in the event of occurrence of loss or damage covered
B·3·3	Restore the damaged work
B·3·4	Authorise the insurers to pay all monies to the employer
C·4·1	Give notice to CA in the event of occurrence of loss or damage covered
C·4·3	Authorise the insurers to pay all monies to the employer
C·4·5	Restore the damaged work
Schedule 4	Endeavour to agree the amount and method of opening up and testing

(cl 2·17·1). Provided the contractor complies with this requirement, it is not liable for any non-compliant work, other than the CDP Works (cl 2·17·3). Once either the contractor or contract administrator discovers a divergence, the contract administrator must issue an instruction within seven days to rectify the situation. This is treated as a variation (cl 2·17·2), and may therefore give rise to an extension of time and reimbursement of direct loss and expense (cl 2·29·1 and 4·24·1).

4.28   The contractor may need to take immediate action in an emergency, but only in so far as is reasonably necessary to comply with statutory requirements (cl 2·18). Unless it is an emergency, any alteration made by a contractor, for example at the request of a district surveyor, could be a breach of contract, unless the contract administrator decides to sanction the variation under clause 3·14·4.

4.29   Under clause 3·25 each party undertakes to the other to comply with all their obligations under the CDM Regulations. Clause 3·25·1 places a contractual obligation on the employer to ensure that the planning supervisor carries out his or her duties under the Regulations. This is a wider obligation than the 'reasonable satisfaction with competence' obligation imposed by the Regulations. Breach of it gives the contractor the right to terminate the contract under clause 8·9·1·4. What is more likely to happen is that the contractor will claim for an extension of time or direct loss and/or expense for breach of clause 3·25, as would constitute a Relevant Event under clause 2·29·6 and a 'matter' under 4·24·5. An example might be where a planning supervisor delays in commenting on a contractor's proposed amendment to the Health and Safety Plan, and progress is thereby delayed.

4.30   Clause 3·25·2 also places a duty on the contractor, if acting as the principal contractor referred to in article 6, to comply with all the relevant duties set out in the Regulations. Breach of this duty is grounds for termination under clause 8·4·1·5. The warning notice still has to be given, and JCT Practice Note 27 suggests that the provision should only be used for situations where the Health and Safety Executive are likely to close the site. Any breach is covered, however, provided termination is not unreasonable or vexatious, and the contract administrator and the employer should consider any breach that might lead to action being

Figure 8 Key powers of the contractor

2·33	Consent to partial possession by employer
3·7·1	Sublet the works, with CA's consent
3·8·2	Add persons to the list of sub-contractors
3·8·3	Carry out or sublet work to be done by a listed sub-contractor
3·10·1	Make a reasonable objection to an instruction
3·13	Request CA to specify provision empowering an instruction
4·12	Submit an application for interim payment
4·14	Suspend performance of its obligations
4·23	Make written application for loss and/or expense suffered
5·3·1	Disagree with the application of the schedule 2 procedure
8·9·1	Give notice to employer specifying defaults
8·9·2	Give notice to employer specifying suspension events
8·9·3	Terminate its employment because of continuation of specified default or suspension event
8·9·4	Terminate its employment because of repeat of specified default or suspension event
8·10·1	Terminate its employment because of employer insolvency
8·11·1	Terminate its employment because of suspension of the works
Schedule 1	
7	Notify CA disagrees with comment on design document, with reasons
Schedule 2	
1·1	Notify CA that information provided is insufficient to prepare quotation
Schedule 3	
B·2·1·2	Take out and maintain a joint names policy if employer defaults
C·3·1·2	Take out and maintain a joint names policy if employer defaults
C·4·4	Terminate its employment by notice
C·4·4·1	Invoke the dispute resolution procedures with respect to the termination

taken against the employer serious. If work needs to be postponed or other instructions given due to a breach of the contract then there should be no entitlement to an extension of time or direct loss and/or expense.

4.31   The contractor is obliged to send any modification of the Health and Safety Plan to the employer (cl 3·25·2). This may occur, for example, due to unexpected site conditions or to changes in detailed design information from nominated sub-contractors, or through the Contractor's Designed Portion information being provided after the Works commence. The contractor should take the cost of developing the Health and Safety Plan into account at tender stage, and no claims can be made for adjusting it to suit the contractor or sub-contractor's working methods. If alterations are needed as a result of an instruction requiring a variation then the costs are included in valuing the variation, and the alterations may be taken into account in assessing an application for extension of time. The contractor is also obliged to provide the planning supervisor with any information he or she reasonably requires for the preparation of the Health and Safety File (see para 6.19).

Other obligations

4.32   In addition to the major obligations outlined above, the contractor also has other obligations arising out of the contract. Most significant of these are in relation to progress and programming, discussed in this chapter, and in regard to insurance matters, discussed in Chapter 9. The contractor's obligations are summarised in Figure 8.

5 Programme

5.1 Most standard form contracts for use on a traditional procurement route require the agreement of dates for the commencement and completion of the work. It is essential that actual dates are given at the time of tendering, and not vague indications such as 'to be agreed' or 'eight weeks after approval by … '. It is important that the contractor knows the time of year it will be carrying out the Works, as the start date and duration will affect the tender figure.

5.2 In the event that work is started without proper agreement over dates, the contract will be subject to the Supply of Goods and Services Act 1982, which states that completion is to be within a reasonable time. If the contractor fails to complete within a reasonable time the employer will be unable to claim liquidated damages, and will have to prove any damages it wishes to recover.

5.3 The SBC05 Contract Particulars require the entry of a 'date of possession' of the site and a 'date for completion' of the Works. If phased working is required, then the Works may be divided into sections (sixth recital), in which case seperate dates of possession and completion must be entered for each section. The contract allows for the employer to defer possession under an optional clause, and for the contract administrator to adjust the dates for completion in certain circumstances. It also allows for the parties to agree an adjustment as a consequence of a variation to the Works (a 'Pre-agreed Adjustment', see paragraph 5.11). For a summary of the key milestones in SBC05 see Figure 9.

Possession by the contractor

5.4 Possession of the site is a fundamental term of the contract. Failure to give contractor possession is a serious breach by the employer, which may amount to repudiation, and therefore give the contractor the right to treat the contract as at an end, or entitle it to terminate its employment (cl 8·9·2). Giving possession of only part of the site could amount to a breach unless this intention has been made clear in the contract documents (*Whittal Builders* v *Chester-le-Street DC*).

Whittal Builders Co. Ltd v Chester-le-Street District Council (1987) 40 BLR 82

Whittal Builders contracted with the council on JCT63 to carry out modernisation work to 90 dwellings. The contract documents did not mention the possibility of phasing but the council gave the contractor possession of the houses in a piecemeal manner. Even though work of this nature was frequently phased, the judge nevertheless found that the employers were in breach of contract for not giving the contractors possession of all 90 dwellings at the start of the contract, and the contractor was entitled to damages.

Figure 9 The SBC05 timeline

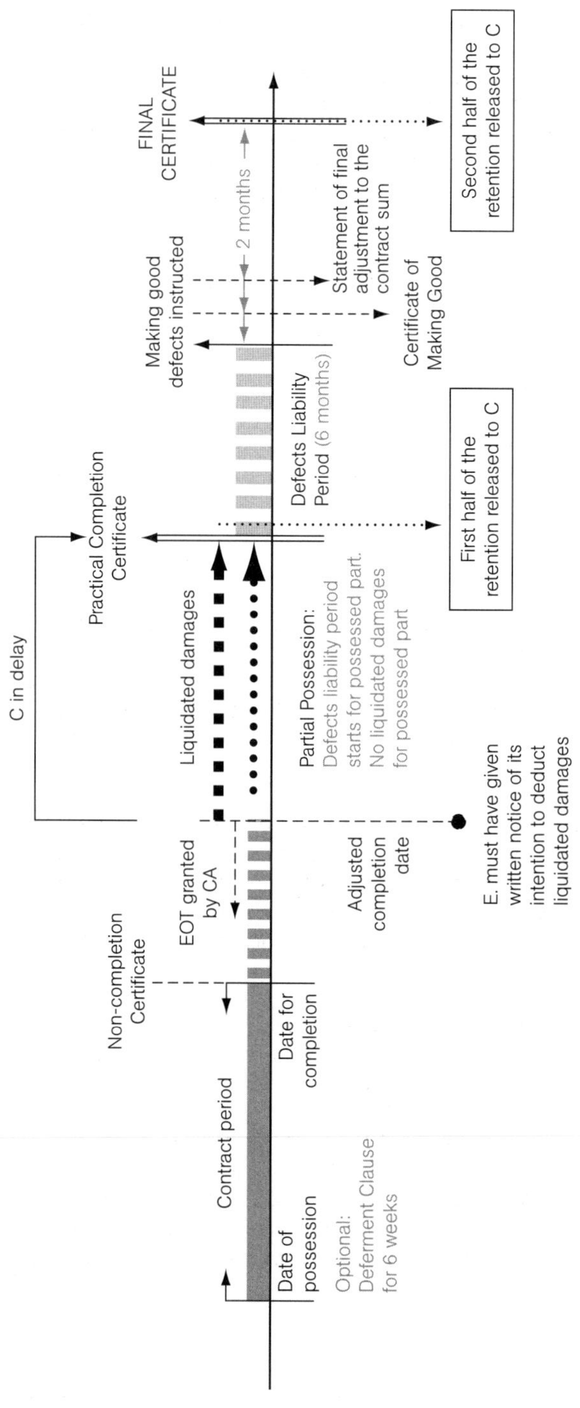

The SBC05 timeline diagram showing: Contract period, Date of possession, Optional: Deferment Clause for 6 weeks, Date for completion, Non-completion Certificate, EOT granted by CA, Adjusted completion date, E. must have given written notice of its intention to deduct liquidated damages, C in delay, Liquidated damages, Practical Completion Certificate, Partial Possession: Defects liability period starts for possessed part. No liquidated damages for possessed part, First half of the retention released to C, Defects Liability Period (6 months), Making good defects instructed, Certificate of Making Good, Statement of final adjustment to the contract sum, 2 months, FINAL CERTIFICATE, Second half of the retention released to C.

KEY:
EOT: Extension of time
CA: Contract Administrator
C: Contractor

Credit: Lucy Murawski

5.5   SBC05 requires that the contractor is given possession of the site on the date of possession (cl 2·4). Under clause 1·1, a Section is defined as a subdivision of the Works, not the site, and clause 2·4 reflects this, stating 'in the case of a Section, possession of the relevant part of the site shall be given'. In some cases, where the Section is all pervasive (for example, where the mechanical and electrical services have been defined as a Section), access to the entire site may be required. Any proposed subdivision of the site in relation to the Sections must be made completely clear in the contract documents, together with proposed arrangements for access to each subdivision.

5.6   The contract allows for the employer to defer the giving of possession of the site or of any relevant part under an option clause (cl 2·5), for a period not exceeding six weeks. In practice this is a very useful provision, as it allows the employer to make small adjustments to the commencement date without having to re-negotiate the contract with the contractor. The tender documents must have stated that the clause is to apply, and the relevant section of the Contract Particulars must be completed. If a period less than six weeks is to be allowed, this must be inserted in the Contract Particulars. Where the Works are split into Sections, it is possible to set different periods of deferment for each Section, again up to a maximum of six weeks.

5.7   If it becomes necessary to defer possession, the employer must notify the contractor in writing. Although the contract does not require it, it would be wise to do this as far in advance of the planned commencement date as possible. The contractor will be entitled to an extension of time (cl 2·29·3) and loss and expense (cl 4·23), and early notification should keep the losses to a minimum. The clause must be operated strictly according to the Conditions, and any delay beyond the periods stated in the Contract Particulars would be a breach of contract.

5.8   The parties are of course always free to renegotiate the terms of any contract. Therefore, if there is a delay in giving possession which is longer than the amount stated in the Contract Particulars, or where the Contract Particulars have stated that clause 2·5 does not apply, the parties may have to agree a new date of possession, usually with a financial compensation to the contractor. Any further delay beyond the agreed date would of course be a breach.

5.9   Degree of possession is such that there must be no interference that prevents the contractor from working in whatever way or sequence it chooses. With most jobs this means that the contractor must be given clear possession of the whole site up until practical completion. Where clear possession is not intended, then the tender documents should set out in detail the restrictions and the contract must be amended accordingly. Should the employer wish to use any part of the Works for any purpose during the time that the contractor has possession, this should also be made clear in the tender documents, otherwise it can only be with the agreement of the contractor (cl 2·6·1). Similarly, should the employer wish the contractor to allow access for others to carry out work, this should also be made clear (cl 2·7·1).

Progress

5.10   It would normally be implied into a construction contract that a contractor will proceed 'regularly and diligently' with the work, and this is an express term in SBC05 (cl 2·4). The contractor is free to organise its own working methods and sequence of operations, with the qualification that it must comply with statutory requirements and the Health and Safety Plan (cl 2·1). This freedom has been held to be the case even where the contractor's chosen sequencing may cause extra cost to the employer with the operation of fluctuation provisions (*GLC* v *Cleveland Bridge and Engineering*).

Greater London Council v Cleveland Bridge and Engineering Co. (1986) 34 BLR 50 (CA)

The GLC employed Cleveland Bridge to fabricate and install gates and gate arms for the Thames Barrier. The specially drafted contract provided dates by which Cleveland Bridge had to complete certain parts of the Works. Clause 51 was a fluctuations provision which allowed for adjustments to be made to the contract sum if, for example, the rates of wages or process of material rose or fell during the course of the contract. The clause also stated 'provided that no account shall be taken of any amount by which any cost incurred by the Contractor has been increased by the default or negligence of the Contractor'. The contract was lengthy, and Cleveland Bridge left a part of the work to be done at the very end of the period, but delivered the gates on time. The result was that the GLC had to pay a large amount of fluctuations in respect of the work done at the last minute. The GLC argued that the contractor had failed to proceed regularly and diligently, and therefore was in default.

The court held that even if the slowness of the contractor's progress might at certain points have given the employer the right to terminate the contract under the termination provisions, this would not by itself be a breach of contract as referred to in clause 51. The contractor could organise the work any way it wished provided it completed on time, and was therefore owed the full amount of the fluctuations. (Note that JCT forms now contain a freezing provision which prevents fluctuations from operating when the contractor is in culpable delay.)

5.11   SBC05 requires the contractor to produce a 'master programme' (cl 2·9·1·2). The programme is not a contract document, and the clause expressly states that it does not impose any additional obligations on either party (for example, it would not oblige the contractor to carry out the work in any particular sequence). The programme may nevertheless be useful to the contract administrator, particularly in regard to monitoring progress and assessing extensions of time. Clause 2·9·1 states that it should be provided 'so soon as is possible' after the execution of the contract, if not previously provided. The contractor is also required to issue 'an amendment or revision of that programme' each time a new completion date is fixed by the contract administrator under clause 2·28·1, or where there is a confirmed acceptance of a Schedule 2 quotation (termed a 'Pre-agreed Adjustment').

5.12   The provisions relating to the programme are brief in comparison with those in many other standard forms, and the employer may wish to consider whether any additional requirements should be included in the tender documents. As in many projects it is important to have a programme at an early stage, and the form has no sanction for its non-provision, it may be wise to insist on the programme at tender stage or before executing the contract. Alternatively, if a more specific time limit was required, for example not less than two weeks before the date for possession, this would require an amendment to clause 2·9·1·2.

5.13   No particular format is required for the programme by SBC05, but again requirements could be set out in the tender documents, for example that it must show a critical path and/or allocation of resources. If the Information Release Schedule is not used it would be open to the contractor to set out dates when key information will be required, although these would not be contractually binding. SBC05 does not require the contractor to publish a revised programme when the contractor is in delay, so if regular updates are required a revision to clause 2·9·1·2 should be considered.

5.14   The contract administrator should be most careful never to 'approve' a programme in such a way that it becomes a contract document against which the contract administrator's own performance in providing information will be judged. A programme which shows a large float period between the contractor's estimated completion date and the date entered in the Contract Particulars should be queried. For example, if the contractor has tendered to carry out the work in 30 months but produces a programme showing estimated completion in 24 months, then the supply of information by the contract administrator might be judged on the 24-month period if the contract administrator appears to have agreed to this. Also, the early completion date may create difficulties for the employer. It is clear, however, that just because a contractor's programme shows an intention to complete early, there is no implied duty on the employer to enable the contractor to achieve this early completion (*Glenlion Construction* v *The Guinness Trust*).

Glenlion Construction Ltd v The Guinness Trust (1987) 39 BLR 89

The Trust employed Glenlion Construction to carry out works in relation to a residential development at Bromley, Kent. The contract was on JCT63, which required the contractor to complete 'on or before' the date for completion, and to provide a programme. Disputes arose which went to arbitration and several questions of law regarding the contractor's programme were subsequently raised in court. The contractor later claimed loss and expense on the ground that it was prevented from economic working and achieving the early completion date shown on its programme only by failure of the contract administrator to provide necessary information and instructions to the dates shown. The court decided that Glenlion was entitled to complete before the date for completion whether or not it was contractually bound to produce a programme and whether or not it did in fact produce one. Glenlion was therefore entitled to carry out the Works in a way which would achieve an earlier completion date. However, there was no implied obligation that the employer (or the contract administrator) should perform its obligations so as to enable the contractor to complete by any earlier completion date shown on the programme.

Completion

5.15   Building contracts normally stipulate a completion date for the Works. The importance of this completion date is that it provides a fixed point from which damages may be payable in the event of non-completion. Generally in construction contracts the damages are 'liquidated', and typically are fixed at a rate per week of over-run.

5.16   The contractor is obliged to complete the Works by the completion date, and in general accepts the risk of all events that might prevent completion by this date. The contractor is relieved of this obligation if the employer causes delays or in some way prevents completion. In addition, most contracts contain provisions allowing for the adjustment of the completion date in the event of certain delays caused by the employer, or neutral delaying events. The contract dates can, of course, always be adjusted by agreement.

5.17   In contracts it is sometimes essential that completion is achieved by a particular date and failure would mean that the result is worthless. This is sometimes referred to as 'time is of the essence'. Breach of such a term would be considered a fundamental breach, and would give the employer the right to terminate performance of the contract, and treat all its own obligations as at an end. The expression 'time is of the essence' is seldom, if ever, applicable to building contracts such as SBC05, as the inclusion of extensions of time and liquidated damages provisions imply that the parties intended otherwise.

5.18   SBC05 uses the following terms:

- 'Date for Completion' – the date inserted in the Contract Particulars, which is the date agreed at the time of entering into the contract. Where the Works are divided into Sections a separate date will be stated for each Section;

- 'Completion Date' – the date for completion of the Works or a Section, or any later date consequent upon an extension of time or a pre-agreed adjustment;

- 'practical completion' – the date at which, in the opinion of the contract administrator, the Works or a Section are complete.

5.19   The form provides for the granting of extensions of time, which results in the fixing of a new completion date (for the Works, or any Section). However the form makes no provision for the contract administrator to reduce the contract period to less than that stated in the Contract Particulars, even when work is omitted (cl 2·28·6·3). Under the schedule 2 quotation procedure (see para 7.14) may agree an extension or reduction to the contract period, termed a 'Pre-agreed Adjustment.'

5.20   If the contractor fails to complete the Works or any Section by the relevant completion date, liquidated damages become payable (see Figure 10).

Extensions of time

Principle

5.21   An important reason for an extension of time clause is to preserve the employer's right to liquidated damages, in the event that the contractor fails to complete on time due in part to some action for which the employer is responsible. If there were no provisions to grant extensions of time, and a delay occurred that was caused at least in part by the employer, this would in effect be a breach of contract by the employer and the contractor would no longer be bound to complete by the completion date (*Peak* v *McKinney*). The employer would therefore lose the right to liquidated damages, even though some of the blame for

Figure 10 Completion and liquidated damages

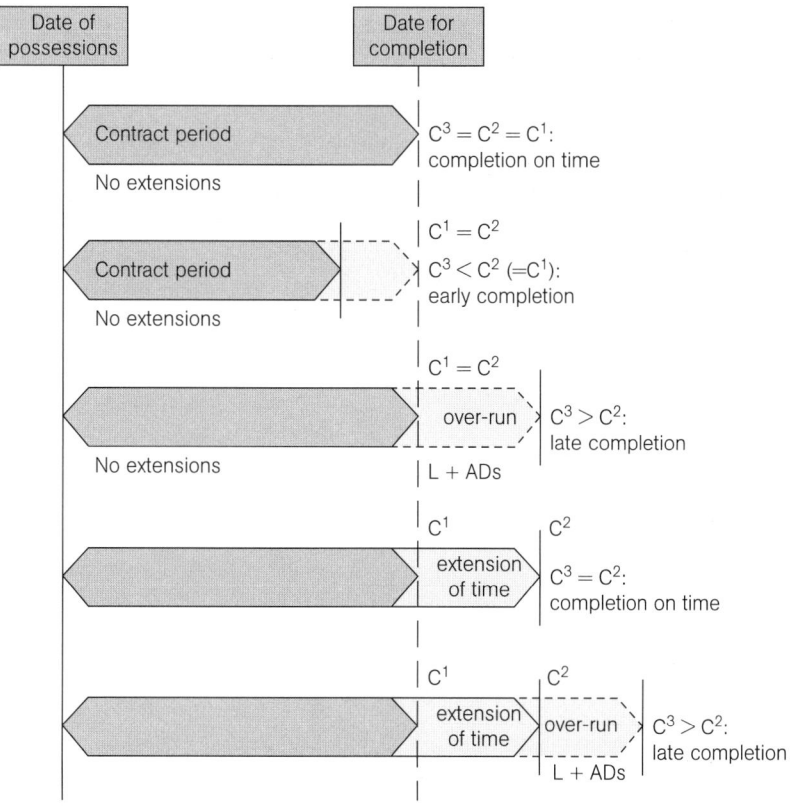

KEY
C1 = Date for completion (entered in Contract Particulars)
C2 = Completion date (as adjusted by extension of time)
C3 = Practical completion (as certified by contract administrator)

the delay may rest with the contractor. On the same principle, where the contract does provide for extending time, but these provisions are not operated, the employer will not be able to levy damages where it is in part responsible for the delay. The phrase 'time at large' is often used to describe these situations. In most cases, however, the contractor would remain under an obligation to complete within a reasonable time.

Peak Construction (Liverpool) Ltd v McKinney Foundations Ltd (1970) 1 BLR 111 (CA)

Peak Construction were main contractors on a contract to construct a multi-storey block of flats for Liverpool Corporation. The main contract was not on any of the standard forms but a contract drawn up by the council. McKinney Foundations were nominated sub-contractors to design and construct the piling. After the piling was complete and the sub-contractors had left the site, serious defects were discovered in one of the piles, and following further investigation minor defects were found in several other piles. Work was halted while the best strategy for remedial work was debated between the parties. The city surveyor did not accept the initial remedial proposals, and it was agreed that an independent engineer would prepare a proposal. The Council refused to agree in advance to accept his decision, and delayed making the appointment. Altogether it was 58 weeks before work resumed, although the remedial work took only six weeks, and the main contractors brought a claim against the sub-contractors for damages.

The Official Referee at first instance found that the entire 58 weeks delay was caused by the nominated sub-contractor and awarded £40,000 damages for breach of contract, based in part on liquidated damages, which the corporation had claimed from the contractor. McKinney appealed, and the Court of Appeal found that the 58 weeks could not possibly be all due to the breach of the sub-contractor, but was in part caused by the tardiness of the Corporation. This being the case, and as there were no provisions in the contract for extending time for their delay, the Corporation lost their right to claim liquidated damages, and this component of the damages awarded against the sub-contractor was disallowed. Even if the contract had contained such a provision, the failure of the contract administrator to exercise it would have prevented the Corporation from claiming liquidated damages. The only remedy would have been for the Corporation to prove what damages they had suffered as a result of the breach.

Procedure

5.22   In SBC05 the provisions for granting an extension of time are set out under clauses 2·26–2·29. The contractor gives written notice 'forthwith' to the contract administrator, when progress to the Works or any Section is being or is likely to be delayed (2·27·1). This appears to require the contractor to keep alert to future probable delays. The notice must be given whether or not completion is likely to be delayed, and whether or not the delay is caused by a Relevant Event (i.e. the requirement to give a notice is not limited to circumstances where the contractor is claiming an extension of time).

5.23   The notice should set out the material circumstances and causes of the delay and identify any Relevant Events (cl 2·27·1). The notice must include or be followed by further particulars in respect of each and every Relevant Event, including the delay caused by each of those events and an estimate by the contractor of its effect on completion (cl 2·27·2). The contractor's notice and particulars, including its estimate, appear to be a condition precedent for the granting of an extension of time. The contract administrator is therefore not obliged to issue an extension until the contractor has properly complied with clause 2·27. The contractor is obliged to keep the contract administrator informed of any changes in the estimated delay and in the particulars provided, and to 'supply such further information as the Architect/Contract Administrator may at any time reasonably require' (cl 2·27·3). It is suggested that the contract administrator's right to require further information is not restricted to any changes in the delay, but would include information relating to any of the matters raised in the notice.

5.24   The amount of detail required is sometimes disputed but in order to give proper consideration to a notice there must be sufficient information. It likely that clause 2·27·2 would be interpreted as an obligation to supply a reasonable level of detail, particularly in respect of information which only the contractor has access to. The contract administrator should ask for more information if this is necessary to make a fair and reasonable assessment. This must never be regarded as a delaying tactic, however. The contract administrator must reach a decision 'on the balance of probability' taking into account all information available, including matters the administrator has knowledge of and matters that can be established through reasonable enquiry. An extension, which later appears insufficient in the light of further information, can always be adjusted at review.

5.25   In regard to Relevant Events the following points should be noted:

- 'Any impediment, action or default … ' (cl 2·29·6) covers a very wide range of possible acts by the employer, contract administrator, quantity surveyor or 'employer's person', and would include, for example, failure to provide a drawing at the time shown on the information release schedule.

- Clause 2·29·7 applies only to Statutory Undertakers operating independently. Work carried out by a Statutory Undertaker under a contract with the employer would fall under clause 2·29·6, and if under a contract with the Contractor, any delays would be at the Contractor's risk.

- 'Weather' is to be exceptional and adverse (i.e. not that which would be expected at the time of year in question) (cl 2·29·8). The effect of the weather is assessed at the time the work is actually carried out, not when it should have been, according to the contractor's programme. The contractor will normally provide weather records to support its claim.

- 'Specified Perils' (cl 2·29·9) can under certain circumstances include events caused by the contractor's own negligence.

- Civil commotion and terrorism under 2·29·10 includes the threat of terrorism, and activities of local authorities in dealing with such threats.

- The very wide protection afforded with respect to strikes, and not simply those directly affecting the Works, but also those causing difficulties in preparation and transportation of goods and materials, or the preparation of the design for the contractor's designed portion. Such strikes will not necessarily be confined to the UK and, given the current extent of overseas imports, the effects could be considerable (cl 2·29·11).

- 'Force majeure' (cl 2·29·13) is a French term used 'with reference to all circumstances independent of the will of man, and which it is not in his power to control'. It includes Acts of God and other matters outside the control of the parties. However, many items under this category, for example strikes, fire and weather, are dealt with elsewhere in the contract.

5.26   The contract administrator must respond 'as soon as is reasonably practicable' and within 12 weeks from receipt of the 'required particulars' (cl 2·28·2) (see Figure 6). It is not clear whether this phrase is referring to the particulars given in the contractor's notice or subsequently (cl 2·27·2), or any particulars requested by the contract administrator (cl 2·27·3). It is likely, however, that time would start to run from the point where sufficient information has been provided, and if this occurred with the first notice then time would run from that date. In any event, the requirement to respond 'as soon as is reasonably practicable' should not be overlooked. What time period is 'reasonably practicable' will depend on the nature of the delay, and the complexity of the information to be assessed, but once sufficient information is available the contract administrator should proceed to making a decision, and not wait until the end of the 12-week period. If the completion date is less than 12 weeks away the contract

Figure 11 Extensions of time

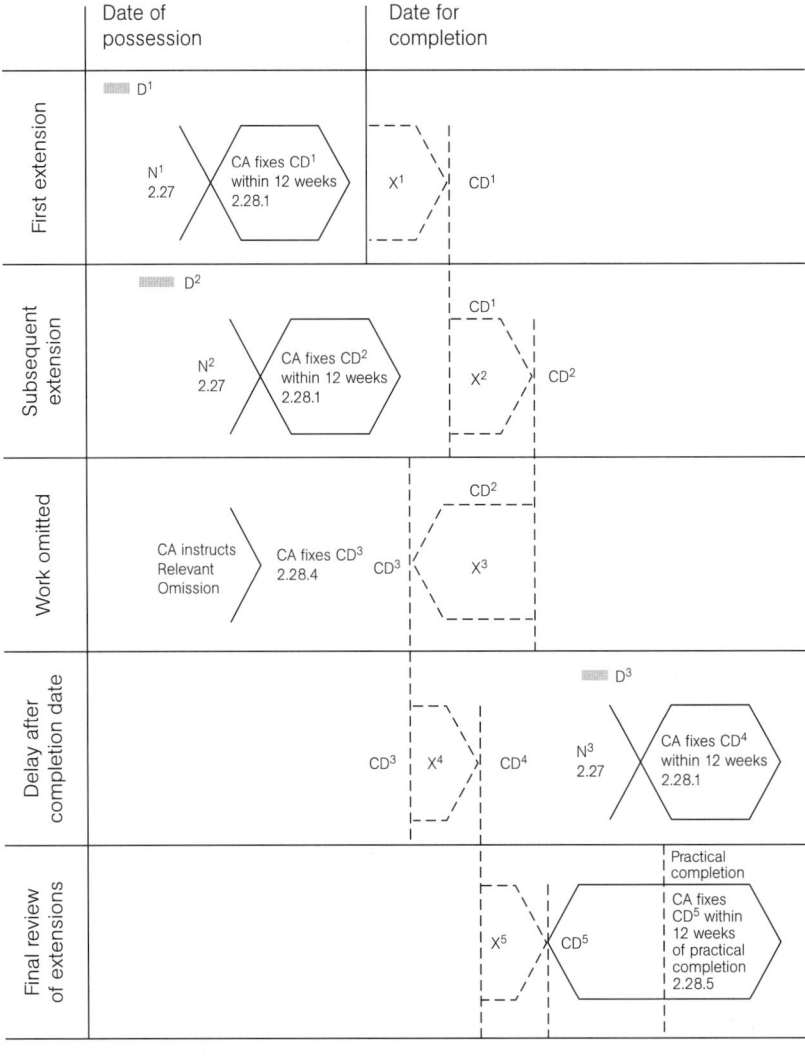

KEY
CA = Contract administrator   CD = Completion date   D = Delaying event
X = Extension of time   N = Notice

administrator must endeavour to respond by the completion date. This is not stated to be an absolute obligation, but as discussed above the contract administrator should proceed as soon as is practicable – any undue delay may result in the contractor claiming that it is no longer bound to finish by the current completion date.

5.27   The contract administrator must either fix a new completion date for the Works or Section, or notify the contractor that no extension of time is due (cl 2·28·2). It is important to note that SBC05 now requires the contract administrator to state in the decision the amount of extension attributed to each relevant event, in other words the contract administrator must apportion the decision (2·28·3·1).

5.28   The contract administrator may reduce a previous extension of time by fixing an earlier completion date having regard to 'Relevant Omissions', i.e. where work has been omitted through a variation instruction (cl 2·28·4), in which case the contractor must be notified of the reduction attributed to each omission (cl 2·28·3·2). An extension or reduction may also result from a 'Pre-agreed Adjustment'. The contract administrator may not, however, fix a date earlier than the date for completion entered in the Contract Particulars (cl 2·28·6·3), nor alter the length of a pre-agreed adjustment unless the work which was the subject of the pre-agreed adjustment is itself subject to a 'Relevant Omission' (cl 2·28·6·4).

5.29   It appears that the contract administrator may not issue an extension of time before the completion date unless a notice of delay under clause 2·27 has been given by the contractor, and may not award one except in respect of Relevant Events identified by the contractor. If the contract administrator becomes aware of such matters he or she could alert the contractor, but would be under no obligation to do so.

5.30   After the completion date of the Works or any Section has passed, the contract administrator may review the extensions of time given, and must do so prior to 12 weeks after practical completion (cl 2·28·5). The review may extend or bring forward the completion date, or confirm the date previously fixed and must be notified to the contractor together with the apportionment required by clause 2·28·3. At this point the contract administrator must take into account Relevant Events not notified by the contractor (cl 2·28·5·1). As above the contract administrator can only reduce the extensions of time already given if a reduction or omission of work instructed after the last completion date was fixed would justify this (cl 2·28·5·2), and may not fix a date earlier than the original 'date for completion'.

5.31   If the contract administrator awards a further extension of time in respect of Relevant Events which occur after the date for completion or any extended completion date, i.e. when the contractor is in 'culpable delay' (*Balfour Beatty* v *Chestermont Properties*), the extension is added onto the date that has passed, referred to as the 'net' method of extension. It should be noted that this only operates under clause 2·28·5 discussed above. Contractors' applications under clause 2·27·1 can only operate to relate to Relevant Events that occur before the practical completion date (implied by the wording of clause 2·28·1·2 'is being or is likely to be').

Balfour Beatty Building Ltd v Chestermont Properties Ltd (1993) 62 BLR 1

In a contract on JCT80 the Works were not completed by the revised completion date and the contract administrator issued a Non-completion Certificate. The contract administrator then issued a series of variation instructions and a further extension of time which had the effect of fixing a completion date two and a half months before the first of the variations instructions. He then issued a further Non-completion Certificate and the employer proceeded to deduct liquidated damages. The contractor took the matter to arbitration and then appealed certain decisions on preliminary questions given by the arbitrator. The court held that the contract administrator's power to grant an extension of time pursuant to clause 25·3·1·1 could only operate in respect of Relevant Events that occurred before the original or the previously fixed completion date, but the power to grant an extension under 25·3·3 applied to any Relevant Event. The contract administrator was right to add the extension of time retrospectively (termed the 'net' method).

Assessment

5.32   It is an obligation on the contract administrator to issue extensions of time when properly due and any failure on the part of the contract administrator to do so is a breach on the part of the employer. The contract administrator has no power to grant extensions of time except for the 'Relevant Events'. In every case the contract administrator should assess the effect of the delay on the contract completion date. The contractor's programme can be used as a guide but is not binding. The effect on progress is assessed in relation to the work being carried out at the time of the delaying event, rather than the work that was programmed to be carried out.

5.33   Clause 2·28·6·1 contains the important proviso that the contractor must use its 'best endeavours' to prevent delay; therefore the contract administrator can assume that the contractor will take steps to minimise the effect of the delay on the completion date. The phrase 'best endeavours' appears to suggest something more than 'reasonable' or 'practicable' but it is unlikely to extend to excessive expenditure. It should also be noted that the contractor is not to be given an extension of time where delays are caused by the defaults set out in clause 2·20, i.e. errors in the Contractor's Proposals, or delays in submitting Design Documents.

5.34   The effects of any delay on completion – taking into account the contractor's 'best endeavours' – are not always easy to predict. The contract administrator is required to reach an opinion, and in doing this the contract administrator owes a duty to both parties to be fair and reasonable (*Sutcliffe* v *Thackrah*). This applies even where the delay has been caused by the contract administrator, for example where the contract administrator has failed to issue drawings within the time limits stipulated in the contract.

5.35   It sometimes happens that two or more delaying events can happen simultaneously, or with some overlap, and this can raise difficult questions with respect to the awarding of extensions of time. In the case of concurrent delays involving two or more 'Relevant Events', it has been customary to grant the extension in respect of the dominant reason. However this approach was considered inappropriate in relation to direct lost and expense (*H. Fairweather & Co.* v *Wandsworth*). The new provisions in clause 2·28·3 make it clear that the contract administrator must apportion the total delay amongst the various contributing causes. (*H. Fairweather & Co.* v *Wandsworth*).

5.36   Where one overlapping delaying event is a Relevant Event and the other is not, in other words one is the employer's risk and the other the contractor's, a difficult question arises as to what extension of time is due. It would seem logical that the contractor should be given an extension of time for the full length of delay caused by the Relevant Event, irrespective of the fact that during the overlap the contractor was also causing delay. Taking any other approach, by for example splitting the overlap period and awarding only half to the contractor, could result in the contractor being subject to liquidated damages for delay partly caused by the employer.

H. Fairweather & Co. Ltd v London Borough of Wandsworth (1987) 39 BLR 106

Fairweather entered into a contract with the London Borough of Wandsworth to erect 478 dwellings. The contract was on JCT63. Pipe Conduits Ltd were nominated sub-contractors for underground heating works. Disputes arose and an arbitrator was appointed who made an interim award. Leave to appeal was given on several questions of law arising out of the award. The arbitrator had found that where a delay occurred which could be ascribed to more than one event, the extension should be granted for the dominant reason. The dominant reason was strikes, and the arbitrator had granted an extension of 81 weeks for this reason, and made it clear that this reason did not carry any right for direct loss and/or expense. The court stated that an extension of time was not a condition precedent for an award of direct loss and/or expense, and that the contractor would be entitled to direct loss and/or expense for other events which had contributed to the delay.

Partial possession

5.37   Possession by the employer of completed parts of the Works (or any Section) ahead of practical completion of the whole Works is provided for under clauses 2·33–2·37. This 'partial possession' requires the agreement of the contractor, which cannot be unreasonably withheld. When practical completion is 'deemed to have occurred' for the 'relevant part' of the Works (cl 2·34), the rectification period for that part is deemed to have commenced (cl 2·34) and the Certificate of Making Good defects has to be issued for that part separately (cl 2·35). However, it would appear that this remains part of the Works, and is still to be included under the Practical Completion Certificate.

5.38   Liquidated damages are reduced by the proportion of the value of the possessed part of the Works to the contract sum (cl 2·37). The effect of clause 4·20·3 is that half of the retention is released for that proportion of the Works. If insurance Option C applies the employer is responsible for insuring the possessed part under Schedule 3 paragraph C·1. If Options A or B apply the employer may wish to consider insuring the part as the contractor's obligation to insure the Works will cease (cl 2·36).

5.39   Apparently no formal advance written notice is required to bring the provision into operation – but it might be wise to provide it. It is possible that taking partial possession could be considered an 'impediment' entitling the contractor to claim an extension of time and direct loss and/or expense (for example, if access to other parts is affected and causes disruption to the programme), therefore it would be wise to clarify the position with the contractor before a decision is made.

5.40   Once the employer has taken possession the contract administrator must 'thereupon' issue a statement to the contractor identifying precisely the extent of the 'relevant part' (cl 2·33) and the date of possession. This should be done with great care, even using a drawing to illustrate the extent, and communicating to the insurers where relevant. The statement must be issued immediately after the part is taken into possession.

5.41   It is important to note that the fact that significant work remains outstanding has not prevented the courts from finding that 'partial possession' has been taken of the whole works, in situations where a tenant has effectively occupied the whole building, allowing access to the contractor for remedial work (see *Skanska Construction (Regions) Ltd* v *Anglo-Amsterdam Corporation Ltd*). If the parties do not intend clause 2·33 to take effect for the whole project, they must make it clear, under a carefully worded agreement what the contractual consequences of any occupation are intended to be (see below).

Skanska Construction (Regions) Ltd v Anglo-Amsterdam Corporation Ltd (2002) 84 ConLR 100

Anglo-Amsterdam Corporation (AA) engaged Skanksa Construction (Skanska) to construct a purpose-built office facility under a JCT81 With Contractor's Design form of contract. Clause 16 had been amended to state that practical completion would not be certified unless the certifier was satisfied that any unfinished works were 'very minimal and of a minor nature and not fundamental to the beneficial occupation of the building'. Clause 17 of the form stated that practical completion would be deemed to have occurred on the date the employer took possession of 'any part or parts of the Works.'

AA wrote to Skanska confirming that the proposed tenant for the building would commence fitting-out Works on the completion date (the air-conditioning system was not functioning and Skanska had failed to produce operating and maintenance manuals). Following this date the tenant took over responsibility for security and insurance, and Skanska was allowed access to complete outstanding work. AA alleged that Skanska was late in the completion of the Works and applied liquidated damages at the rate of £20,000 per week for a period of approximately nine weeks. Skanska argued that the building had achieved practical completion on time or that alternatively partial possession of the Works had taken place and that in consequence its liability to pay liquidated damages ceased under clause 17.

The matter went to arbitration and subsequently Skanska appealed it to the court. The court was also unhappy with the decision. The court found that clause 17·1 could operate equally when possession had been taken of all parts of the Works and was not limited to possession of only part or some parts of the Works. Accordingly, it found that partial possession of the entirety of the Works had in fact been taken some two months earlier than the date of practical completion, as a result of AA agreeing to the tenant commencing fit-out Works. Consequently, even though significant Works remained outstanding, Skanska was entitled to repayment of the liquidated damages that had already been deducted by AA.

Use or occupation before practical completion

5.42   When the contractor is still in possession of the site, and the employer wishes to 'use or occupy the site or the Works or part of them', then clause 2·6 provides for this. The purposes for which the employer might require this are described simply as 'storage or otherwise', so in theory at least the clause places no limits on what form the use might take. The consent of the contractor in writing is required, and the contractor would not be unreasonable in withholding permission unless the intervention and inconvenience are likely to be minimal, or unless an agreement is reached over granting an extension of time and/or compensating the contractor for any losses. Before the contractor is required to give consent, the employer or the contractor as appropriate must notify the insurers (cl 2·6·1). If under Option A any additional premium is required, the contractor notifies the employer of the amount and this is added to the contract sum (cl 2·6·2).

5.43   Towards the end of a project the situation often arises where the contractor has not completed by the date for completion and, although no sections of the Works are sufficiently complete to allow the employer to take possession of those parts under clauses 2·33–2·37, the employer is nevertheless anxious to occupy at least part of the Works. Clause 2·6 appears to cover early occupation, but great care must be taken to agree to all the contractual consequences before the occupation takes place. A suggestion was put forward in the 'Practice' section of the RIBA Journal (February 1992) which is frequently found useful in practice (see Figure 12). In this arrangement, in return for being allowed to occupy the premises, the employer agrees not to claim liquidated damages during the period of occupation. Practical completion obviously cannot be certified, and there is no release of retention money until it is. Matters of insuring the Works will need to be settled with the insurers.

5.44   Because such an arrangement would be outside the terms of the contract it should be covered by a properly drafted agreement which is signed by both parties. (For illustrations of the importance of drafting a clear agreement see *Skanska* v *Anglo-Amsterdam Corporation* above and *Impresa a Castelli* v *Cola*.) It may also be sensible to agree that in the event that the contractor still fails to achieve practical completion by the end of an agreed period, liquidated damages would run again, possibly at a reduced rate. In most circumstances this arrangement would be of benefit to both parties, and is far preferable to issuing a heavily qualified Practical Completion Certificate listing 'except for' items.

Impresa a Castelli SpA v Cola Holdings Ltd (2002) CLJ 45

Impressa agreed to build a large four-star hotel for Cola Holdings Ltd (Cola), on the basis of the JCT Standard Form of Building Contract With Contractor's Design, 1981 edition. The contract provided that the Works would be complete within 19 months from the date of possession. As the work progressed it became clear that the completion date of February 1999 was not going to be met, and the parties agreed a new date for completion in May 1999 (with the bedrooms available to Cola in March) and a new liquidated damages provision of £10,000 per day as opposed to the original rate of £5,000. Following the agreement, further difficulties with progress were encountered which meant that the May 1999 completion date was also not going to be met. The parties entered into a second variation agreement, which recorded that access for Cola would be allowed to parts of the hotel to enable it to be fully operational by September 1999 despite certain Works not being complete (including the air conditioning). In September 1999, parts of the hotel were handed over but Cola claimed that such parts were not properly completed. A third variation agreement was set up with a new date for practical completion and for the imposition of liquidated damages.

Disputes arose and amongst other matters Cola claimed for an entitlement for liquidated damages. Impressa argued that it had achieved partial possession of the greater part of the works, resulting in a reduced rate of liquidated damages per day being payable. The court found that although each

Figure 12 Practice Section, *RIBA Journal* (February 1992)

Contracts

**Employer's possession
before practical completion
under JCT contracts**

It is not uncommon for the Employer after the completion date has passed to wish to take possession of the Works before the Contractor has achieved practical completion. In this event an ad hoc agreement between Employer and Contractor is required to deal with the situation.

In respect of such an agreement members may wish to have regard to the following note …

Outstanding items

Where it is known to the architect that there are outstanding items, practical completion should not be certified without specially agreed arrangements between the employer and the contractor. For example, in the case of a contract where the contract completion date has passed it could be so agreed that the incomplete building will be taken over for occupation subject to postponing the release of retentions and the beginning of the defects liability period until the outstanding items referred to in a list to be prepared by the architect have been completed, but relieving the contractor from liability for liquidated damages for delay as from the date of occupation, and making any necessary changes in the insurance arrangements. In such circumstances either the Certificate of Practical Completion form should be used for it should be altered to state or refer to the specially agreed arrangements. In making such arrangements the architect should have the authority of the client-employer.

When the Employer is pressing for premature practical completion there is a need to be particularly careful where there are others who are entitled to rely on the issue of a Practical Completion Certificate and its consequences. In the case where part only of the Works is ready for hand-over in partial possession provisions* can be operated to enable the Employer with the consent of the Contractor to take possession of the completed part.

* JCT 80: clause 18
JCT 84: clause 2.11 in appendix to JCT Practice Note IN/1 (if incorporated in the contract)

---

variation agreement could have used the words 'partial possession' they had in fact instead used the word 'access'. The court could not find anything in the variation agreements to suggest that partial possession had occurred. It said that the natural conclusion from this was that the agreements related to use and occupation as referred to in clause 23·3·2 of the contract, and the agreed liquidated damages provision was therefore enforceable.

Practical completion

5.45   The contract administrator is obliged to certify practical completion of the Works or a Section (cl 2·30) when in the contract administrator's opinion the following three criteria are fulfilled:

- Practical completion of the Works or the Section is achieved (see below);

- the contractor has complied 'sufficiently' with clause 2·40 (supply of as-built drawings in relation to the Contractor's Designed Portion);

- the contractor has complied 'sufficiently' with clause 3·25·3 (supply of information required for the Health and Safety File).

5.46   The wording of clause 2·30 is not as clear as it might be, in that the first part appears to be stating that the contract administrator is certifying three separate things: 'Practical Completion of the Works' plus the two others listed above. However, it must be remembered that this wording is the result of two later insertions being placed in a clause which originally referred solely to practical completion of the Works. It is suggested that a correct analysis of this clause is that practical completion only occurs when all three conditions are met, the principal argument for this being that only one date is entered on the certificate. This should be the date when the last condition is fulfilled, in other words, if there is a delay before receiving the 'as built' drawings, the date of their receipt should be the date on the certificate, irrespective of the fact that practical completion of the Works was achieved days or even weeks earlier.

5.47   It should be noted that the contractor's obligation to supply information for the Health and Safety File depends on the planning supervisor having requested it in writing. Even then the use of the term 'sufficiently complied' may allow the contract administrator to use his or her discretion in issuing the certificate with some information missing. The contract administrator should, however, be very careful not to place the employer in a position where it would be in breach of the CDM Regulations.

5.48   Deciding when the Works have reached practical completion often causes some difficulty. In the leading commentary, *Keating on Building Contracts* (May 1995) the editors submit that the following is the correct analysis:

(a) the Works can be practically complete notwithstanding that there are latent defects;

(b) a Practical Completion Certificate may not be issued if there are patent defects. The Defects Liability Period is provided in order to enable defects not apparent at the date of practical completion to be remedied (*Jarvis & Sons* v *Westminster Corporation* and *H. W. Neville (Sunblest)* v *William Press*);

(c) Practical completion means the completion of all the construction that has to be done (*Jarvis & Sons* v *Westminster Corporation*);

(d) however the contract administrator is given a discretion under clause 17·1 to certify practical completion where there are very minor items of work left incomplete, on 'de minimus' principles (*H. W. Neville (Sunblest)* v *William Press*). (*Keating*, at page 673)

City of Westminster v J. Jarvis & Sons Ltd (1970) 7 BLR 64 (HL)

Jarvis entered into a contract with the City of Westminster to construct a multi-storey car park in Rochester Row. The contract was on JCT63 which included 'delay on the part of a nominated sub-contractor… which the Contractor has taken all reasonable steps to avoid' (cl 23(g)). The piling was carried out by nominated sub-contractors who completed their work by the date required under their sub-contract and withdrew from site. Subsequently, defects were discovered in many of the piles and the remedial works caused a delay of over 21 weeks to the contractor. The House of Lords found that on a proper interpretation of clause 23(g) delay on the part of the nominated sub-contractor only occurred if it had failed to complete its work by the date in the sub-contract. The clause did not apply after the Works had been accepted as complete. In addressing the question of whether there was a delay in completion the court also had to consider what was meant by 'practical completion' of the sub-contract Works. Viscount Dilhourne stated:

> The contract does not define what is meant by 'practically completed'. One would normally say that a task was practically completed when it was almost but not entirely finished; but 'practical completion' suggests that that is not the intended meaning and that what is meant is the completion of all the construction work that has to be done. (at page 75)

H. W. Neville (Sunblest) Ltd v William Press & Son Ltd (1981) 20 BLR 78

William Press entered into a contract with Sunblest to carry out foundations, groundworks and drainage for a new bakery on a JCT63 contract. A Practical Completion Certificate was issued, and new contractors commenced a separate contract to construct the bakery. A Certificate of Making Good Defects and a Final Certificate were then issued for the first contract, following which it was discovered that the drains and the hard standing were defective. William Press returned to site and remedied the defects, but the second contract was delayed by four weeks and Sunblest suffered damages as a result. It commenced proceedings, claiming that William Press was in breach of contract and in their defence William Press argued that the plaintiffs were precluded from bringing the claim by the conclusive effect of the Final Certificate. Judge Newey decided that the Final Certificate did not act as a bar to claims for consequential loss. In reaching this decision he considered the meaning and effect of the Practical Completion Certificate and stated:

> I think that the word 'practically' in clause 15(1) gave the contract administrator a discretion to certify that William Press had fulfilled its obligation under clause 21(1) where very minor de-minimus work had not been carried out, but that if there were any patent defects in what William Press had done then the contract administrator could not have issued a Practical Completion Certificate. (at page 87)

5.49   In *Keating on Building Contracts* it is also advised that the discretion of the contract administrator should be exercised with caution, and this is undoubtedly sound advice. The decision as to when practical completion has occurred is one of the most critical decisions that the contract administrator has to make in administering the contract, as the consequences to the employer are significant. The author frequently encounters instances of contract administrators certifying practical completion qualified by long lists of 'snagging' items. Even though the employer, anxious to move into the newly completed works, may initially agree to the early certification, the following contractual problems will remain unresolved:

- half of the retention will be released for that section or the whole works, leaving only 1.5 per cent retention in hand (cl 4·20·3). This puts the employer at considerable risk, as the 1.5 per cent is only intended to cover latent, not patent, defects;

- the relevant rectification period begins (cl 2·38);

- the onus shifts to the contract administrator to instruct all necessary outstanding work under clause 2·38·1. If the contract administrator fails to instruct something, the contractor would have no authority to enter the site to carry it out – therefore the contract administrator will inevitably become involved in managing and programming the outstanding work;

- the contractor's liability for negligent damage to the Works commences (cl 6·3·1·1);

- the contractor's duty to insure the Works under Option A ends;

- the contractor's liability for liquidated damages ends (cl 2·32·2);

- regular Interim Certificates cease to be issued (cl 4·9·2);

- the employer will be the 'occupier' for the purposes of the Occupiers Liability Act and also may be subject to claims regarding health and safety.

5.50   A certificate must be issued as soon as the criteria in clause 2·30 are met for each section (termed a 'Section Completion Certificate') and for the whole Works (the 'Practical Completion Certificate'). The contractor is obliged to complete 'on or before' the completion date and once practical completion is certified the employer is obliged to accept the Works. Employers who wish to accept the Works only on the date in the contract would need to amend the wording.

Procedure at practical completion

5.51   The contract sets out no procedure for what happens at practical completion, it simply requires the contract administrator to certify it. The contract bills may set out a procedure, and the contract administrator should check carefully at tender stage to ensure that the procedure is satisfactory. In particular, the specification or bills may stipulate commissioning procedures for mechanical and electrical services, and testing procedures to demonstrate compliance with building regulations.

5.52   Leading up to practical completion it appears to be widespread practice for contract administrators to issue 'snagging' lists, sometimes in great detail and on a room-by-room basis. The contract does not require this, and neither do most standard terms of appointment. Under the contract, responsibility for quality control and snagging rests entirely with the contractor. In adopting this role the contract administrator may be assisting the contractor, and although this may appear to benefit the employer it may lead to confusion over the liability position, which could cause problems at a future date.

5.53   It is frequently the practice for the contractor to arrange a 'handover' meeting. The term is not used in SBC05 and although handover meetings can be of use, particularly in introducing the finished project to the employer, it may be better not to set out complex or inflexible procedures in the bills of quantities. Even where a handover meeting has been arranged, or the contractor has stated in writing that the Works are complete, it remains the contract administrator's responsibility to decide when practical completion has been achieved. If the contract administrator feels that the Works are not complete there is no obligation to justify this opinion with schedules of outstanding items. It is suggested that the best course may be to draw attention to typical items, but to make it clear that the list is indicative and not comprehensive.

Failure to complete by the completion date

5.54   In the event of failure to complete a Section or the Works by the relevant date, the contract administrator is required to certify this fact by means of a 'Non-completion Certificate' (cl 2·31). As this certificate is a condition precedent to deduction of liquidated damages (cl 2·32·1·1), it is important that it is issued promptly. Once the certificate has been issued, the contractor is said to be in 'culpable delay'. The employer, provided that it has issued the necessary notices (see paragraph 5.57), may then deduct the damages from the next Interim Certificate, or reclaim the sum as a debt. Note that fluctuations provisions are frozen from this point. If a new completion date is later set this has the effect of cancelling the Non-completion Certificate and no additional written cancellation is needed (cl 2·32·3).

Liquidated and ascertained damages

5.55   The agreed rates for liquidated and ascertained damages for each Section, or for the Works is entered in the Contract Particulars. This is normally expressed as a specific sum per week (or other unit) of delay, to be allowed by the contractor in the event of failure to complete by the completion date (note there may be several different rates where the Works are divided into Sections). The amount must be calculated on the basis of a genuine pre-estimate of the loss likely to be suffered. Provided that it is, the sum will be recoverable without the need to prove the actual loss suffered, and irrespective of whether the actual loss is significantly less or more than the recoverable sum. In other words, once the rate has been agreed, both parties are bound by it. If 'nil' is inserted then this may preclude the employer from claiming any damages at all (*Temloc* v *Erril*), whereas if the Contract Particulars are left blank the employer may be able to claim general damages.

Temloc Ltd v Erril Properties (1987) 39 BLR 30 (CA)

Temloc entered into a contract with Erril Properties to construct a development near Plymouth. The contract was on JCT80 and was in the value of £840,000. '£ Nil' was entered in the Contract Particulars against clause 24·2, liquidated and ascertained damages. Practical completion was certified around six weeks later than the revised date for completion. Temloc brought a claim against Erril Properties for non-payment of some certified amounts, and Erril counter-claimed damages for late completion. It was held by the court that the effect of '£ nil' was not that the clause should be disregarded (because, for example, it indicated that it had not been possible to assess a rate in advance), but that it had been agreed that there should be no damages for late completion. Clause 24 is an exhaustive remedy and covers all losses normally attributable to a failure to complete on time. The defendant could not, therefore, fall back on the common law remedy of general damages for breach of contract.

5.56 Before liquidated and ascertained damages may be claimed, the following pre-conditions must have been met:

- the contractor must have failed to complete the Works by the completion date;

- the contract administrator must have fulfilled all duties with respect to the award of an extension of time;

- the contract administrator must have issued a Non-completion Certificate (cl 2·32·1·1);

- the employer must have informed the contractor before the date of the Final Certificate, that it may require the payment of liquidated damages, or deduct liquidated damages from monies due (cl 2·32·1·2).

5.57 If these conditions are fulfilled, then the employer may give notice to the contractor in writing that it requires the contractor to pay liquidated damages, or that it intends to withhold or deduct the damages from monies due (cl 2·32·2). In both cases the notice should state whether the contract rate or a lesser rate will be applied. The notification must be given to the contractor no later than five days before the final date for payment of the debt due under the Final Certificate. The requirement and notification must be reasonably clear but there is no need for a great deal of detail (*Finnegan* v *Community Housing Association*). It may be possible for the clause 2·32·1·2 'requirement in writing' and the clause 2·32·2 notice to be dealt with together, provided the document contains the necessary information and is issued at the right time. Similarly, if the employer wishes to deduct liquidated damages from an amount payable on a certificate, then a further, separate notice will be needed, unless the clause 2·32·1 notice complies with the contractual requirements for withholding notices (see Footnote 38 of SBC05 and paragraph 8.25 of this Guide).

J. F. Finnegan Ltd v Community Housing Association Ltd (1995) 77 BLR 22 (CA)

Finnegan were employed by the Housing Association to build 18 flats at Coram Street, West London. The contractor failed to complete the work on time, and the contract administrator issued a Certificate of Non-completion. Following the Practical Completion Certificate an Interim Certificate was issued. The employer sent a notice at the same time as the cheque honouring the certificate, which gave minimal information (i.e. not indicating how LADs had been calculated). The Court of Appeal considered this sufficient to satisfy the requirement for the employer's written notice in clause 24·2·1. Peter Gibson LJ stated (at page 33):

> I consider that there are only two matters which must be contained in the written requirement. One is whether the employer is claiming a payment or a deduction in respect of LADs. The other is whether the requirement relates to the whole or a part (and, if so, what part) of the sum for the LADs.

He then stated (at page 35):

> I would be reluctant to import into this commercial agreement technical requirements which may be desirable but which are not required by the language of the clause and are not absolutely necessary.

The requirements as to notices have now changed. However, there appears to be no reason why the general comments would not still apply, i.e. that the amount of information required would be no more than the minimum set out in the contractual provisions.

5.58   If an extension of time is given following the issue of a Non-completion Certificate then this has the effect of cancelling that certificate. A new Non-completion Certificate must be issued if the contractor then fails to complete by the new completion date (cl 2·31). The contract states that the employer does not need to inform the contractor again that they may claim damages, as clause 2·32·1·2 remains satisfied (cl 2·32·4). The contract does not address the question of whether a new clause 2·32·2 notice is required, but for the sake of clarity it may be wise to do this.

5.59   Once the completion date is adjusted the employer must, if necessary, repay any liquidated damages recovered for the period up to the new completion date (cl 2·32·3). In *Department of Environment for N. Ireland* v *Farrans* it was decided that the contractor has the right to interest on any re-paid liquidated damages. This decision, however, was on JCT63 and, given that the SBC05 clauses now expressly refer to repayment without stipulating that interest is due, it would appear that interest is not now due in this situation. This was the view taken by His Honour Judge Carr in the first instance decision of *Finnegan* v *Community Housing Association* (1993) 65 BLR 103 (at page 114).

Department of Environment for Northern Ireland v Farrans (Construction) Ltd (1981) 19 BLR 1 (NI)

Farrans was employed to build an office block under JCT63. The original date for completion was 24 May 1975 but this was subsequently extended to 3 November 1977. During the course of the contract the contract administrator issued four Certificates of Non-completion. By 18 July 1977 the employer had deducted £197,000 in liquidated damages but following the second Non-completion Certificate re-paid £77,900 of those deductions. This process was repeated following the issue of the subsequent Non-completion Certificates. Farrans brought proceedings in the High Court of Justice in Northern Ireland, claiming interest on the sums that had been subsequently re-paid. The court found for the contractor, stating that the employer had been in breach of contract in deducting monies on the basis of the first, second and third certificates, and that the contractor was entitled to interest as a result. The BLR commentary should be noted, which questions whether a deduction of liquidated damages empowered by clause 24·2 can retrospectively be considered a breach of contract, but the case has not been over-ruled.

5.60   Certificates should always show the full amount due to the contractor. It is the employer alone that makes the deduction of liquidated damages. The employer would not be considered to have waived its claim by a failure to deduct damages from the first or any certificate under which this could validly be done, and would always be able to reclaim them as a debt at any point up until the Final Certificate.

6   Control of the Works

6.1   The contract administrator's duties to the employer are normally set out in an appointment document, frequently on a standard form produced by one of the professional institutions. The contract administrator's role within the construction contract, however, is to be determined solely from the wording of the form. A contract administrator named in SBC05 will be required to take certain actions, for example supplying necessary information, issuing instructions and issuing certificates or statements. In some matters the contract administrator will act as agent of the employer, for example when issuing instructions which vary the Works, and in others as an independent decision maker, for example when issuing certificates or deciding on claims for direct loss and/or expense. Failure by the contract administrator to comply with any obligation (usually prefaced by the phrase 'the contract administrator shall') will constitute failure on the part of the employer tantamount to breach of contract (see Figures 13 and 14).

6.2   Direct control over the carrying out of the contract Works, including the manner in which the Works are undertaken, is, however, solely the responsibility of the main contractor (cl 3·6). The duty of the contract administrator to the employer will normally be to inspect the work at intervals. The contractor is obliged to provide the contract administrator with reasonable access to the Works, its premises, and to sub-contractors premises (cl 3·1). The precise obligation and purpose of such visits will arise directly from the terms (either express or implied) of the professional appointment, and of course SBC05 includes no express provision relating to inspection or monitoring of work by the contract administrator. Clearly when the contract administrator is required under the contract to form an opinion on various matters, including the standard of work and materials prior to issuing a certificate, then it would be implied, even if not expressly set out in the terms of appointment, that some form of inspection must take place.

Person in charge

6.3   The contractor is required to keep constantly on the site a competent 'person-in-charge' (cl 3·2) who receives any instructions given by the contract administrator or directions given by the clerk of works, and therefore acts as the contractor's agent on site. Although there is no requirement in the contract Conditions to have the person named, it would be sensible to agree who the person will be in a pre-contract meeting and record this in the minutes. The contract administrator has the power to exclude persons from site (cl 3·21), but may not do so 'unreasonably or vexatiously'. This power is only likely to be used where it appears that the employee may be seriously affecting operations on site. If used unreasonably, this could constitute an 'impediment' under clauses 2·29·6 and 4·24·5.

Employer's representative

6.4   Although the majority of the administration of the contract is dealt with by the contract administrator, the employer has an active role to play and is required to make decisions on

Figure 13 Key powers of the contract administrator

2·20·2	Apply in writing to contractor for design documents
2·24	Consent to removal of goods etc. from site
2·28·4	Fix an earlier completion date
2·28·5	Following practical completion, fix an earlier or later completion date
2·38·2	Issue instructions requiring defects to be made good
3·4	Confirm in writing clerk of works directions
3·8·2	Add persons to the list of sub-contractors
3·11	Issue notice requiring compliance with an instruction
3·14·1	Issue instructions requiring a variation
3·14·4	Sanction in writing any variation made by the contractor
3·15	Issue instructions postponing work
3·17	Issue instructions requiring inspections or tests
3·18·1	Issue instructions requiring removal of work from site
3·18·2	Allow non-compliant work etc. to remain
3·18·3	Issue instructions requiring variation resulting from non-compliant work
3·18·4	Issue instructions requiring further inspections or tests
3·19	Issue instructions in relation to manner of carrying out work, including compliance with Health and Safety Plan
3·21	Issue instructions excluding employed persons from site
5·3·1	Require a schedule 2 quotation
8·4·1	Give notice to contractor specifying defaults

various matters, and to issue notices direct to the contractor, and is entitled to exercise various powers (see Figures 15 and 16). The employer is entitled to appoint a representative to exercise all these functions (cl 3·3). The contractor must be notified in writing of the identity of the individual, and of any exceptions to the functions the individual will perform. Footnote 39 to the form makes it clear that to avoid any confusion in the roles, neither the contract administrator nor the quantity surveyor should be appointed as the employer's representative.

Clerk of works

6.5   The employer is entitled to employ an independent clerk of works whose duty is 'to act solely as an inspector on behalf of the Employer under the direction of the Contract Administrator' (cl 3·4).

Figure 14 Key duties of the contract administrator

1·9	Certificates to be issued to employer with copy to contractor
2·8·2	Provide contractor with copies of contract documents
2·9·1·1	Provide contractor with copies of schedules
2·10	Determine levels and provide contractor with setting out drawings
2·11	Provide contractor with copies of information on information release schedule
2·12·1	Provide contractor with copies of 'reasonably necessary' information
2·16	Notify contractor of agreement or decision regarding CDP discrepancy
2·15	Issue instruction in relation to discrepancies
2·17·1	Notify the contractor of any discrepancy between documents and statutory requirements
2·17·2	Issue instructions relating to 2·17·1 discrepancy
2·28·1	Give extensions of time
2·28·2	Notify contractor of decision regarding extensions of time
2·28·3	Apportion extension of time between relevant events
2·30	Issue a Practical Completion Certificate for the works or a section
2·31	Issue a Non-completion Certificate for the works or a section
2·38·1	Issue a schedule of defects that appear during the rectification period
2·40	Issue the Certificate of Making Good
3·12	Issue all instructions in writing
3·16	Issue instructions in relation to provisional sums
3·18·2	Confirm decision to allow non-compliant work etc. to remain in writing
3·20	Give reasons for dissatisfaction with work within a reasonable time
3·23	Issue instructions regarding antiquities
3·24	Ascertain amount of loss and/or expense suffered due to discovery of antiquities
4·5·2·1	Ascertain amount of loss and/or expense due
4·5·2	Send copy of statement of final adjustment prepared by quantity surveyor to contractor
4·9·1	Issue Interim Certificates
4·15·1	Issue Final Certificate
4·18·2	Prepare a statement specifying amount of retention deducted
4·19·1	Prepare a statement specifying amount of retention that would have been deducted

68   **Guide to SBC05**

Figure 14 Continued

4·23	Ascertain amounts of loss and/or expense
8·7·4	Issue a certificate setting out an account of balance due (unless employer issues a statement)
Schedule 1	
2	Return a copy of each design document submitted marked A, B or C
4	Identify why CA considers design document is not in accordance with the contract
7	Confirm or withdraw comment following notification by contractor
Schedule 2	
1·1	Supply contractor with further information to prepare quotation
4	Instruct whether or not the variation is to be carried out
Schedule 3	
A·4·4	Issue certificates to cover insurance monies paid
Schedule 4	Endeavour to agree the amount and method of opening up and testing, consider the stated criteria when issuing instructions

Figure 15 Key duties of the employer

2·4	Give possession of the site
3·5·1	Nominate a replacement CA
3·8·3	Add persons to the list of sub-contractors
3·25	Comply with CDM Regulations
3·25·1	Ensure planning supervisor and principal contractor (if not the contractor) comply with CDM Regulations
3·25·2	Notify CA and planning supervisor of changes to H&S Plan notified by contractor
3·26	Notify the contractor of any appointment of a replacement planning supervisor or principal contractor
4·6·1	Pay VAT properly chargeable
4·13·3	Notify contractor of amount of payment proposed to be made
4·13·4	Notify contractor of any amount to be deducted
4·13·5	Pay contractor amount specified in notices
4·13·6	Pay interest to the contractor on unpaid amounts

Figure 15 Continued

4·15·3	Notify contractor of amount of payment proposed to be made
4·15·4	Notify contractor of any amount to be deducted
4·15·5	Pay contractor amount specified in notices
4·15·6	Pay interest to the contractor on unpaid amounts
4·18·3	Place retention in a separate banking account, and certify that this has been done
4·18·4	Inform contractor of amount of withholding by reference to latest statement of retention
4·19·3	Release retention deducted because of failure of contractor to provide bond
4·19·4	Deduct amount not covered by bond as retention
6·9·1	Ensure joint names policy provides for recognition of, or waives any right of subrogation against, any sub-contractor
6·10·1	Notify the contractor if notified that terrorism cover has ceased
6·10·2	Notify the contractor whether employment is to continue or terminate
6·14	Comply with the Joint Fire Code, and ensure all employer's persons comply
8·7·4	Issue a statement setting out an account of balance due (unless CA issues a certificate)
8·7·5	Pay the contractor any balance due
8·8·1	Notify the contractor of decision not to complete the works, send the contractor a statement of the balance due
8·10·2	Give the contractor notice if it makes any proposal, etc., in relation to insolvency
8·12·3	Prepare an account (if contractor not required to do so)
8·12·5	Pay the contractor the amount properly due
Schedule 2	
1·1	Notify the contractor that it wishes to accept a schedule 2 quotation
Schedule 3	
A·4·4	Pay all insurance monies to the contractor under certificates
B·1	Take out and maintain a joint names policy for full reinstatement value of the works
B·2·1·1	Produce documentary evidence and receipts
B·2·1·2	Produce a copy of the cover certificate
C·1	Take out and maintain a joint names policy in respect of existing structures
C·2	Take out and maintain a joint names policy for full reinstatement value of the works
C·3·1·1	Produce documentary evidence and receipts
C·3·1·2	Produce a copy of the cover certificate

Figure 16 Key powers of the employer

2·5	Defer possession of the site
2·6·1	Use or occupy the site
2·7·2	Have work executed by employer's persons
2·32	Notify the contractor of intention to deduct liquidated damages
2·33	Take possession of part of the works prior to practical completion, with contractor's consent
3·3	Appoint a representative, terminate such appointment and appoint a replacement
3·4	Appoint a clerk of works
3·8·2	Add persons to the list of sub-contractors
3·11	Employ and pay others to carry out work
4·13·2	Exercise right to withhold sums from monies due
6·4·3	Take out 6·4 insurance if contractor defaults and deduct amount from monies due
6·10·2	Terminate the contractor's employment
7·2	Assign the right to bring proceedings to any transferee
7A·1	Give notice stating that third party rights shall vest in a purchaser or tenant
7B·1	Give notice stating that third party rights shall vest in a funder
7C	Give notice requiring contractor to enter into a warranty with a purchaser or tenant
7D	Give notice requiring contractor to enter into a warranty with a funder
7E	Give notice requiring contractor to comply with requirements set out in contract particulars as to obtaining sub-contractor warranties with purchasers and tenants/funders
7F	Give notice requiring contractor to comply with requirements set out in contract particulars as to obtaining sub-contractor warranties with the employer
8·4·2	Terminate the contractor's employment because of continuation of specified default
8·4·3	Terminate the contractor's employment because of repeat of specified default
8·5·1	Terminate the contractor's employment because of contractor insolvency
8·5·3·3	Take reasonable measures to ensure that the site etc. is protected
8·6	Terminate the contractor's employment because of corruption
8·7·1	Employ and pay other persons to carry out and complete the works, enter upon the site and use temporary buildings, etc.
8·7·5	Pay the contractor any balance due
8·11·1	Terminate the contractor's employment because of suspension of the works

Figure 16 Continued

9·4·1	Give notice of arbitration
9·4·2	Give further notice of arbitration
9·7	Apply to the courts to determine a question of law
Schedule 2	
1·1	Notify the contractor that it wishes to accept a schedule 2 quotation
Schedule 3	
A·2	Take out and maintain a joint names policy if contractor is in default, deduct amounts payable from monies due
A·3	Inspect the contractor's policy and premium receipts or require they are sent to the architect
Schedule 3	
A·4·45	Retain amounts to cover professional fees from insurance monies
A·5·2	Instruct contractor not to renew terrorism cover (if a local authority)
C·4·4	Terminate the contractor's employment by notice
C·4·4·1	Invoke the dispute resolution procedures with respect to the termination

6.6   It should be noted that the presence of a clerk of works does not lessen the contract administrator's duty in respect of site inspection (*Kensington Health Authority* v *Wettern*). The clerk of works does not act as agent for the contract administrator but may issue directions to the contractor. However, any such direction must be one which the contract administrator could have made under the contract and must be confirmed in writing by the contract administrator within two working days if it is to be effective. The contractor will frequently take the initiative and ask for an instruction to confirm the clerk of works' direction, but there is no requirement for the contractor to do this, and therefore the contract administrator and clerk of works should be careful to coordinate over such matters.

Kensington and Chelsea and Westminster Area Health Authority v Wettern Composites (1984) 31 BLR 57

Wettern Composites were sub-contractors for the supply and erection of pre-cast concrete mullions for an extension to the Westminster Hospital, on which the Health Authority had also engaged contract administrators, engineers and a clerk of works. Tersons Ltd were the main contractors. The hospital was completed in 1965 and in 1976 it was discovered that there were considerable defects in the mullions. The Health Authority brought an action against the contract administrators, engineers and sub-contractors, though the latter subsequently went into liquidation. Judgment was given for the Authority. The contract administrators had failed to exercise reasonable skill and care in ensuring conformity of the Works to the design. Although a clerk of works was employed, this did not lessen the contract administrator's responsibility. However, the Health Authority was vicariously liable for the contributory negligence of their clerk of works, and the damages recoverable from the contract administrators were reduced by 20 per cent accordingly.

Planning supervisor

6.7   The planning supervisor's duties derive from the CDM Regulations, and it is the employer's obligation under the contract to ensure that the planning supervisor complies with these (cl 3·25·1). It is the contractor's responsibility to develop the Health and Safety Plan so that it complies with the Regulations, and to ensure that the Works are carried out in accordance with the Plan. The planning supervisor will monitor the development of the Plan, and will coordinate the health and safety aspects of any design changes, but has no duty to inspect the Works and would be unlikely to visit the site unless there is some very unusual circumstance, such as the discovery of an unanticipated hazard. The main responsibility for ensuring correct health and safety measures are employed on site rests with the contractor.

Information to be provided by the contract administrator

6.8   In the majority of construction contracts the information contained in the contract documents will not be sufficient to enable the project to be constructed. Even if the Works have been fully specified it is likely, for example, that information regarding assembly, location, detail dimensions, colours, etc. will be needed by the contractor throughout the project. Supply of this information will usually form part of the contract administrator's duties to the employer under the terms of appointment.

6.9   The SBC05 refers in four places to the contract administrator's obligation to provide information. These refer to 'descriptive schedules or similar documents' (cl 2·9·1·1); setting out information (cl 2·10); 'information referred to in the Information Release Schedule' (cl 2·11); and 'such further drawings or details which are reasonably necessary to explain and amplify the Contract Drawings' (cl 2·12·1). Although the clauses do not require this information to be released under a contract administrator's instruction, this is frequent practice and is wise as it would enable the clause 3·11 provisions to be brought into operation if necessary (see paragraph 6.30 on contract administrator's instructions below). If any of the information supplied introduces changes or additions to the Works it must be covered by a contract administrator's instruction requiring a variation.

6.10   The 'descriptive schedules or other like documents' are to be provided as soon as possible after the execution of the contract (cl 2·9·1·1). These appear to be by way of amplification of that which was given in the tender documentation, as the clause makes it clear that nothing in these documents 'shall impose any obligation beyond those imposed by the contract documents' (cl 2·9·1).

6.11   Under clause 2·10 the contract administrator is responsible for giving sufficient accurately dimensioned drawings and levels to enable the contractor to set out the Works, and the contractor must accurately follow this information. The contractor must 'amend any errors' that result from its own inaccurate setting out. Alternatively, the contract administrator, with the employer's consent, may instruct that the error remains, in which case 'an appropriate deduction ... shall be made from the Contract Sum' (cl 2·10). There is no suggestion in the Conditions as to how this might be assessed, and in practice it will be a matter for negotiation. The error and the deduction should first be discussed with the employer, and the agreed deduction should ensure adequate compensation.

6.12   Information shown on the Information Release Schedule must be supplied at the stipulated date, unless the contract administrator is prevented from doing so 'by an act or default of the Contractor' (cl 2·11). Failure to provide the information and instructions under clause 2·11 would constitute a 'Relevant Event' under clause 2·29·6 and a 'matter' which may give rise to a direct loss and/or expense claim under clause 4·24·5, and possible grounds for termination under clause 8·9, but only where such failure has led to the suspension of the carrying out of the whole of the Works for at least a period which has been entered in the Contract Particulars (cl 8·9·2). An act or default of the contractor might include, for example, failure to provide design documents as required by the contract and which the contract administrator needs to finalise part of the design.

6.13   There is no mechanism whereby the contract administrator may unilaterally adjust the Schedule following an extension of time, for example. Such adjustments will have to be negotiated and agreed by the parties, and it may be necessary to do this on a regular basis, keeping the contract administrator involved (the contract allows the employer and contractor to agree changes to the Information Release Schedule under clause 2·11). Parties should tackle this cooperatively, and note that the contract states such agreement should not be unreasonably withheld. For example, if variations have been issued that involve additional work and have resulted in an extension of time, or if work has been omitted and an earlier completion date fixed, then it would be reasonable for the Schedule to be adjusted to reflect this. If the contractor refuses to agree to an adjustment, the document will become worthless with respect to assessing extensions of time.

6.14   With respect to information not shown on the Schedule, or where a Schedule is not used, the contract administrator is under an obligation to provide 'such further drawings and details as are reasonably necessary either to explain and amplify the Contract Drawings ... and shall issue such instructions ... as are necessary to enable the contractor to carry out and complete the Works' (cl 2·12·1). It is suggested that this obligation would extend to both amplification of information in the Contract Documents and providing full information regarding any variation that is required to be carried out. The inclusion of the word 'reasonably' suggests that the contractor can be expected to obtain some detailed information, for example manufacturers' fixing information. The contract administrator should be careful, however, in respect of leaving decisions to the contractor, as it may not always be possible to hold the contractor responsible should a detail or fixing fail.

6.15   The information and instruction should be provided in sufficient time to allow the contractor to complete by the completion date or, if the contractor appears unlikely to complete by this date, at a date when 'having regard to the progress of the Works' it is reasonably necessary for the contractor to receive the information (2·12·2). There is no general requirement for the contractor to apply for information in writing, but if the contractor has 'reason to believe the contract administrator is not aware' of when information may be needed, the contractor should advise the contract administrator (2·12·3). In practice, such advice may frequently be in the form of a programme indicating dates when information is required. As above, failure to provide the information and instructions under clause 2·12·1 would constitute a 'Relevant Event' under clause 2·29·6 and a 'matter' which may give rise to a direct loss and/or expense claim under clause 4·24·5, and possible grounds for termination under clause 8·9·2.

6.16   Whether or not an Information Release Schedule is used, if any acceleration to the Works is proposed (for example under the acceptance of a Schedule 2 quotation) the contract administrator should take care to warn the client at an early stage if this might present difficulties for programming of information.

6.17   Under SBC05 there is no provision for other consultants issuing information direct to the contractor; this would have to be done through the contract administrator. Delay in supplying necessary drawings by other consultants would therefore have the effect under the contract of a delay on the part of the contract administrator, i.e. a delay for which the employer is responsible. An obligation to supply information on time would normally be implied into the terms of engagement of any consultant, if not expressly set out (Royal Brompton Hospital v Frederick Alexander Hammond).

Royal Brompton Hospital National Health Trust v Frederick Alexander Hammond and others (No 4) [2000] BLR 75

The Royal Brompton Hospital (RBH) engaged Frederick Alexander Hammond to undertake a £19 million construction project on a JCT80 standard form of contract. The contractor successfully claimed against RBH, including for losses suffered due to delays. RBH commenced proceedings against 16 defendants, who were all members of the professional team. A trial was fixed to deal with a number of different issues, all of which were settled except for one relating to the consulting M&E engineers, Austen Associates Ltd. The issue was whether AA were obliged to provide coordination and builder's work information so as to ensure that RBH complied with clause 5·4 of the main contract. The court decided that AA were under a duty to use reasonable skill and care to ensure that the drawings were provided in time to enable the contractor to prepare his installation drawings, and thus to carry out and complete the Works in accordance with the contract Conditions.

Information provided by the contractor

6.18   The contractor may be required to provide information in regard to CDM Regulation requirements, and in relation to the completion of design for which the contractor is responsible, under the Contractor's Designed Portion of the Works.

6.19   In relation to the CDM Regulations, the contractor as 'Principal Contractor' may be required by the planning supervisor to provide information in relation to the Health and Safety File (cl 3·25·3). The contractor is obliged to provide any information the planning supervisor 'reasonably requires', within any time period reasonably required, provided the planning supervisor notifies the contractor in writing. In addition, the contractor as a 'designer' under

Regulation 13 of the CDM Regulations would have a statutory duty to provide information to the planning supervisor. This role is expressly acknowledged under clause 2·9·2.

Contractor's Designed Portion

6.20 In addition to the obligations in relation to the CDM Regulations, SBC05 contains detailed provision regarding the submission of the developing design by the contractor. This information is essential in order for the contract administrator and employer to monitor the development of the design and to integrate it with the rest of the Works.

6.21 The contractor must provide two copies of:

> such Contractor's Design Documents and (if requested) related calculations and information, as are reasonably necessary to explain or to amplify the Contractor's Proposals; and ... all levels and setting out dimensions which the Contractor prepares or uses for the purposes of carrying out and completing the Contractor's Designed Portion. (cl 2·9·2)

6.22 'Design Documents' are defined as 'the drawings, details and specification of goods and workmanship and other related documents prepared by or for the Contractor in relation to the Contractor's Designed Portion'. The information is to be provided 'as and when necessary from time to time in accordance with the Contractor's Design Submission Procedure ... ', and 'the Contractor shall not commence any work to which such a document relates until the Procedure has been complied with ...' (cl 2·9·3). Clause 1 of Schedule 2 additionally requires that the information is submitted in sufficient time for the contract administrator to be able to comment and for the comments to be incorporated.

6.23 In practice there could be differences of opinion as to what information may be 'reasonably necessary'. The information the contractor may need to actually construct the work may be different from the information that the employer and contract administrator would like to receive. Differences could also arise regarding the appropriate timing of submission of this information.

6.24 The form therefore anticipates that the employer will wish to set out its own particular requirements for submissions. Clause 2·9·3 states the information is to be provided 'in accordance with the Contractor's Design Submission Procedure ... or as otherwise stated in the Contract Documents'. The Design Submission Procedure is set out in Schedule 1, which states that the Documents should be submitted 'in such a format as is stated in the Employer's Requirements' and also refers to 'the period of submission... stated in the Contract Documents' (Schedule 1:1). It would therefore be open to the employer, and on most projects would be wise, to include detailed requirements regarding scope, format and timing of submissions in the contract documents.

6.25 Following submission of a design document, the contract administrator must respond within 14 days of the date of receipt, 'or (if later) 14 days from either the date or expiry of the period for submission of the same stated in the Contract Documents' (Schedule 1:2) (in other words, if the contractor supplies information earlier than any agreed date, the contract administrator would not be required to respond any earlier than stated in the contract documents).

6.26   The contract administrator is entitled to take three alternative courses of action (see Figure 8); it can accept the design document, in which case it should return it marked 'A Action'. It may accept it, subject to certain comments being incorporated, in which case it should be marked 'B Action'. Or it can make comments and require the contractor to re-submit the document with the comments incorporated for further approval, in which case it should be marked 'C Action' (Schedule 1:5). In the cases of B Action or C Action the contract administrator must state why the document does not comply with the contract (comments are only valid if the document does not comply (Schedule 1:4) – if it does comply, any required alteration would constitute a variation). If the contract administrator does not respond within the specified period, it is deemed to have accepted the document (Schedule 1:3).

6.27   Schedule 1:8·3 states that no comments or any action by the contract administrator will relieve the contractor from its liability to ensure that the document complies with the contract, or that the project complies with the contract. This has the effect that if the contractor incorporates a comment made by the contract administrator then it accepts that the comment has been properly made (i.e. it identifies a way in which the design document is not in accordance with the contract).

6.28   If the contractor disagrees with a comment and considers that the document complies with the contract, it is required to inform the contract administrator, within seven days of receipt of the comment, that compliance with the comment would give rise to a variation (Schedule 1:7). The contractor must give reasons as to why it thinks this. The contract administrator must either confirm or withdraw the comment within seven days. The confirmation or withdrawal is stated not to signify that the employer accepts that the design document complies, or that the comment represents a variation (Schedule 1:8·1) – this would be a question of fact, if necessary to be resolved by adjudication. The contractor would have to implement the comment and argue its case later.

6.29   If the contractor does not notify the contract administrator of its disagreement with a comment, then the comments will not be treated as a variation, even if it could be later shown in fact to be a variation (Schedule 1:8·2).

Contract administrator's instructions

6.30   Only the contract administrator is given the power to issue instructions (cl 3·10). Sometimes the contract administrator may issue instructions (for example instructions requiring a variation under clause 3·14), but at other times the contract administrator shall issue instructions (for example instructions regarding discrepancies between contract documents under clause 2·15). The latter is an obligation. If the employer gives an instruction other than through the contract administrator this would not be effective under the contract. The contractor would be under no obligation to comply with any such instruction. If the contractor, however, does carry out the instruction a court might consider that there had been an agreed amendment to the contract. The consequences of such an agreement

Figure 17 Procedure for submission of Design Documents

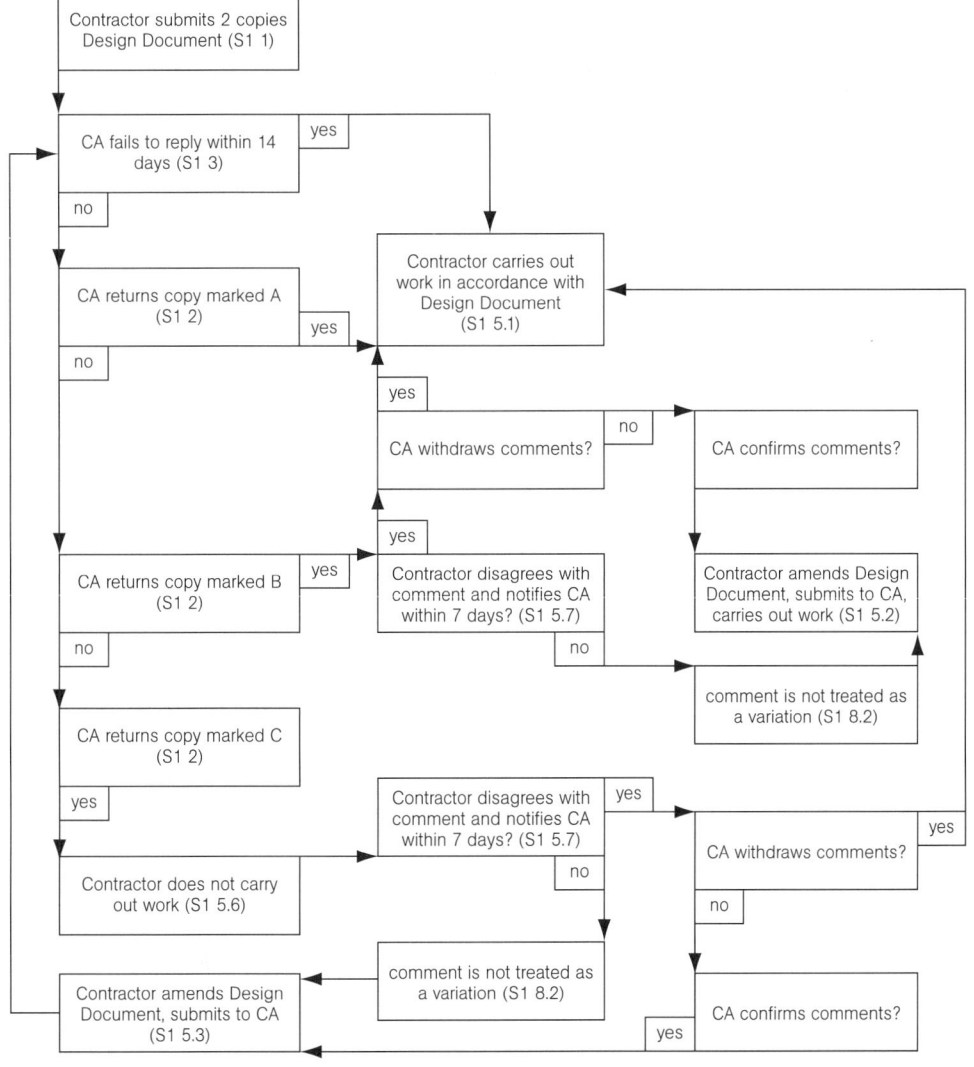

would be difficult to sort out in practice and the employer would be very unwise to make such agreements or issue any instructions other than through the contract administrator.

6.31   Clause 3·12·1 states that all instructions must be in writing. It then states that if any instruction is given 'otherwise than in writing' (which would normally mean oral instructions) it is of no immediate effect, but the contractor must confirm the instruction in writing within seven days (cl 3·12·2). It then takes effect after seven days have elapsed from the date of the contractor's confirmation, provided that the contract administrator has not, in the meantime, dissented in writing. The contract administrator must therefore check carefully that the confirmation represents exactly what is required. Of course the contract administrator may choose to confirm the instruction, in which case it takes effect from the date of the contract administrator's confirmation. If the contractor carries out the instruction on the basis of an oral instruction only, then the contract administrator can later sanction the instruction at any time prior to the issue of the Final Certificate (cl 3·12·4), but the contractor would be taking a risk (*MOD* v *Scott Wilson Kirkpatrick and Partners [2000] BLR*).

MOD v Scott Wilson Kirkpatrick and Partners [2000] BLR (CA)

SWK were engaged as structural engineers and supervising officer by the MOD in relation to refurbishment of the roof at Plymouth Dockyard under GCWks/1. Several years after the Works were completed, wind lifted a large section of roof and deposited it in a nearby playing field. The contract required nails 9" to 12", but the contractor used 4" nails. The supervising officer had been party to discussions regarding using the 4" nails, but neither he nor the contractor could remember very clearly when these discussions happened, or exactly what was said. The Court of Appeal decided the evidence was sparse and vague, and declined to find that there was any instruction under 7(1)(a) or 7(1) (m) (instructions that may be given orally), or any agreement as to the replacement. Even if the supervising officer's conduct amounted to confirmation or encouragement, this could not absolve the contractor from its duty to fix the purlins in a workmanlike manner. The MOD were therefore not precluded from insisting on their strict contractual rights. The Court of Appeal noted, however, that an instruction in writing was not a condition precedent to a claim by the contractor, had they been able to prove that the change had been agreed.

6.32   There is no special format required for instructions but it is often convenient to use the forms published by RIBA Publishing. Instructions in site meeting minutes may constitute a written confirmation of an oral instruction if issued by the contract administrator, but only a 'confirmation' if issued by the contractor, which would only take effect seven days after the date of the minutes. In both cases it would depend on the circumstances whether the minutes were sufficiently clear to fall within the terms of the contract, and it is therefore not good practice to rely on this method.

6.33   The contractor must comply with every instruction (see Figure 18) provided that it is valid, i.e. provided that it is in respect of a matter regarding which the contract administrator is empowered to issue instructions. The contractor must 'forthwith' comply, which for practical purposes means as soon as is reasonably possible (cl 3·10).

6.34   There are three exceptions to the obligation to comply forthwith with a contract administrator's instructions. First, the contractor need not comply with a 5·1·2 instruction (access and use of the site, etc.) to the extent that it makes a reasonable objection (cl 3·10·1). Second, the contractor need not comply with an instruction relating to a Schedule 2 quotation until a confirmed acceptance has been given (cl 3·10·2). Third, the contractor need not comply where it might affect the efficacy of its design (cl 3·10·3, see paragraph 6.42 below).

6.35   If the contractor feels that a contract administrator's instruction might not be empowered by the contract, or requires clarification, then the contractor may ask the contract

Figure 18 Matters about which the contract administrator is empowered to issue instructions

CA may/shall issue instructions:	Clause	
relating to 1·15 discrepancies	2·15	duty
relating to 2·17·1 discrepancies (statutory requirements)	2·17·2	duty
requiring defects to be made good	2·38·2	power
confirming clerk of works directions	3·4	power
varying the works, etc.	3·14·1	power
postponing work	3·15	power
relating to provisional sums	3·16	duty
requiring inspections or tests	3·17	power
requiring removal of work from site	3·18·1	power
requiring a variation resulting from non-compliant work	3·18·3	power
requiring further inspections or tests	3·18·4	power
relating to the manner of carrying out work, including compliance with Health and Safety Plan	3·19	power
excluding employed persons from site	3·21	power
relating to antiquities	3·23	duty
confirming whether or not a Schedule 2 variation is to be carried out	Schedule 2: 4	power

administrator to specify in writing the provisions of the contract under which the instruction is given, and the contract administrator must do this 'forthwith' (cl 3·13). The contractor must then either comply or issue a notice disputing the matter. If, however, the contractor chooses to accept the contract administrator's reply and complies with the instruction, then the employer is bound by the instruction. This would appear to be the case even if at a later stage it is established that the contract administrator had no authority under the contract.

6.36 Even if the contractor decides to query the instruction under 3·13, this does not relieve the contractor from the obligation to comply. Should the instruction be found to be valid the contractor would be liable for any delay caused by failing to comply as required by the contract. If the contractor does comply, but the instruction turns out to have been invalid, the contractor may be entitled to any losses incurred through compliance.

6.37 If the contractor does not comply with a written instruction, the employer may employ and pay others to carry out the work to the extent necessary to give effect to the instruction (cl 3·11). The contract administrator must have given written notice to the contractor requiring compliance with the instruction, and seven days must have elapsed after the contractor's receipt of the notice before the employer may bring in others. This suggests that some recorded form of delivery is desirable.

Variations

6.38 Contract administrator's instructions often require some variation to the Works. Under common law neither party to a contract has the power to unilaterally alter any of its terms.

Therefore, in a construction contract neither the employer nor contract administrator would have the power to require any variations unless the contract contains such a power. As some aspects of construction may be difficult to define exactly in advance, most construction contracts contain provisions allowing the employer to vary the works to some degree. Changes can arise because of a variety of reasons, including unexpected site problems, or because of design changes wanted by the employer, or which become necessary in order to integrate the Contractor's Designed Portion.

6.39   Under SBC05 the contract administrator is empowered to order specific variations (cl 3·14·1). The scope of what constitutes a variation is set out in clause 5·1. It is broadly defined and includes alterations to the quantity and specification of the Works, and to operational restrictions such as access to the site. The contract expressly states that no variation will vitiate the contract (cl 3·14·5) but the power does not extend to altering the nature of the contract, nor can the contract administrator issue variations after practical completion. All variations under clause 3·14 may result in an adjustment of the contract sum (cl 5·6·1) and give rise to a claim for an extension of time (cl 2·29·1) or direct loss and/or expense (cl 4·24·1). If the Works are suspended as a consequence, the variation may also be grounds for termination by the contractor unless the variation is necessitated by some negligence or default of the contractor (cl 8·9·2).

6.40   The contract administrator may vary the Works (cl 5·1·1), for example by changing the standard of a material specified. The contract administrator may add to or omit work, or substitute one type of work for another or remove work already carried out. The contract administrator may vary the access to or use of the site, limitations on working space or working hours, the order in which the work is to be carried out, or any restrictions already imposed (cl 5·1·2). As well as giving the employer a great deal of flexibility, this contract provision is necessary to accommodate difficulties that may arise, for example, through local authority restrictions on working hours. However, the contractor need not comply with a clause 5·1·2 instruction to the extent that it makes reasonable objection (cl 3·14·2). Given that the contractor will be paid for such variations, it is difficult to see what might constitute a 'reasonable' objection but, for example, a variation might have a detrimental 'knock-on' effect on some other project, causing the contractor to suffer losses for which it would not otherwise be compensated.

6.41   Finally, the contract administrator may sanction any variation made by the contractor other than under an instruction of the contract administrator (cl 3·14·4). If such a variation were likely to affect the employer, the contract administrator would be wise to discuss it with the employer before taking action.

Variations to the Contractor's Designed Portion

6.42   Clause 3·14·3 states that where instructions require a Variation in respect of the Contractor's Designed Portion, any instruction 'shall be an alteration to or modification of the Employer's Requirements'. This would appear to prevent the contract administrator from directly requiring changes to the Proposals after the contract is entered into, including to any further design details that are developed as the contract progresses, except in cases where the developing design does not meet the Employer's Requirements. If the contract administrator issues any instruction which in the opinion of the contractor may affect the efficacy of the design, the contractor must object within seven days of receiving the instruction. The instruction will not then take effect until confirmed by the contract administrator (cl 3·10·3).

Defective Work

6.43   Clause 2·3·1 states that all materials, goods and workmanship shall be of the standard specified in the contract documents (see discussion under paragraphs 4.15–4.20). The contract administrator will normally inspect at regular intervals to monitor the standard that is being achieved. However achieving the contractual standard is the responsibility of the contractor, and the lack of an inspection cannot be used as an excuse for sub-standard work. When the standard achieved appears to be unsatisfactory it can be tempting to become involved in directing the day-to-day activities of the contractor on site. Apart from being an enormous burden on the contract administrator, this could confuse the issue of who is ultimately responsible for quality and is to be avoided.

6.44   Where the standard of goods or materials is a matter for the approval of the contract administrator under clause 2·3, any dissatisfaction should be expressed within a reasonable time (cl 3·20). The contract administrator must also state reasons for the dissatisfaction. In all other cases there is no obligation to point out defects or errors, but the contract administrator would normally, of course, draw the contractor's attention to areas of defective or poor quality work. The fact that work has been included in an Interim Certificate does not relieve the contractor from its responsibility for the standard of work, or prevent the contract administrator from deciding that the work is defective (cl 3·6).

Testing work

6.45   The contract administrator may instruct the contractor to open up completed work for inspection, or arrange for testing of any of the work or materials, fixed or unfixed (cl 3·17). No time limit is specified but obviously the contract administrator should instruct as soon as the need for such action becomes apparent (delay could result in escalating or unnecessary costs). Failure to ask for tests in no way relieves the contractor from the obligation to provide work according to the contract. The cost of carrying out the tests is added to the contract sum, unless it was already provided for in the bills of quantities under a provisional sum or unless the work proves to be defective. Unless the work is defective the contractor may also be entitled to an extension of time under clause 2·29·2 and loss and/or expense under clause 4·24·2·2 unless these tests were provided for in the contract bills.

6.46   The contract administrator has several courses of action if work is defective. An instruction can be issued requiring the removal of work, materials or goods from the site (cl 3·18·1); the work, materials or goods can be allowed to remain (cl 3·18·2) (unless part of the Contractor's Designed Portion); a variation can be issued (cl 3·18·3) or further tests can be instructed 'having due regard' to the Code of Practice for tests set out in Schedule 4 (cl 3·18·4).

6.47   If a notice requiring compliance with a clause 3·18·1 instruction to remove from the site is given and not complied with, then the provisions of clause 3·11 could be brought into operation (*Bath and NE Somerset DC* v *Mowlem*). To fall under clause 3·18·1 the instruction must specifically require removal of the work from site, however impractical. Simply drawing attention to the defective work would not be sufficient (*Holland Hannen* v *Welsh Health Technical Services*). Refusal to remove defective work is also a ground for termination under clause 8·4·1·3 provided there has been a written notice or instruction and the refusal materially affects the work.

Bath and North East Somerset DC v Mowlem plc [2004] BLR 153

Mowlem were engaged on JCT98 (Local Authorities With Quantities) to undertake the Bath Spa project. Completion was expected to be in 2002 but work was still underway in 2003. Paint applied by Mowlem to the four pools began to peel, and the contract administrator issued AI Number 103 which required Mowlem to strip and repaint the affected areas. Mowlem refused to comply and the Council issued a notice under clause 4·1·2. Mowlem still did not comply, and the employer engaged Warings to carry out this work. Mowlem refused Warings access to the site, and the Council applied to the court for an injunction, which was granted. Mowlem appealed against the injunction, but the appeal was dismissed.

Mowlem had argued that all the defects could be sorted out by them and that the liquidated damages provided under the contract were the agreed remedy for delays caused. The Council were able to show that the liquidated damages did not adequately compensate them for the losses being suffered. Lord Justice Mance held that in such cases the court should examine whether the liquidated damages would provide adequate compensation, and as in this case they would not, it was appropriate to grant an injunction. In reaching this decision he took into account irrecoverable losses such as the 'unquantifiable and uncompensatable damage to the Council's general public aims'.

Holland Hannen & Cubitts (Northern) Ltd v Welsh Health Technical Services Organisation (1985) 35 BLR 1 (CA)

Cubitts were employed by the Welsh Health Technical Services Organisation (WHTSO) to construct two hospitals at Rhyl and Gurnos. Percy Thomas (PTP) were the contract administrators. Redpath Dorman Long Ltd (RDP) were nominated sub-contractors for the design and supply of pre-cast concrete floor slabs. RDP assured WHTSO that the floors would be designed to CP 116 (concerning deflection), but the design team later required RDP to work to CP 204. Following installation, the contractor complained about extra work and costs due to adjustments to the partitions necessitated by excessive deflection of the floors, and it was established that they had been designed to CP 116 not CP 204.

PTP sent three letters 'condemning' the floors, but the first did not mention clause 6(4), and none of them required removal of the work. Cubitts stopped work for 20 weeks until PTP issued instructions specifying how the defect should be resolved. Cubitts commenced proceedings, claiming compensation for delay. The claim was settled but the relevant parties maintained their proceedings against each other for contribution. The official referee decided that RDL were liable for two-thirds of the amount paid to Cubitts and the design team for one-third. The Court of Appeal decided that this was incorrect and the correct apportionment should have been for RDL to be liable for one-third and the design team for two-thirds. In reaching this conclusion it stated: 'PTP contributed very substantially to the delay which occurred, in failing to recognise the defect in the design at an earlier stage; by issuing an invalid notice in 1976, and by moving very slowly thereafter to take the necessary steps to have the defects in the flooring put right' (Robert Goff LJ).

6.48   If defective work is to be allowed to remain, there must be consultation with the contractor and the approval of the employer must be obtained (cl 3·18·2). The contract administrator must specify in writing exactly which work may remain, and an appropriate deduction is made from the contract sum. Sometimes when defective work is retained, a variation is needed to other work, in order to accommodate the change. If such a consequential variation becomes necessary (again following consultation with the contractor) no addition is made to the contract sum and no extension of time or direct loss and/or expense is given in respect of this (cl 3·18·3). Where the work is part of the Contractor's Designed Portion, the contract administrator has no authority to allow or require it to remain. If the employer preferred the work to stay (perhaps seeking to avoid delays to the programme) this would have to be agreed with the contractor, which would presumably seek to limit its liability for any defects in the non-compliant work.

6.49   If work has been shown to be defective, and further similar non-compliance is suspected, then further tests may have to be ordered (cl 3·18·4). The costs of these, including any direct loss and/or expense, are borne by the contractor. If the work is shown to have been in accordance with the contract, the contractor may be entitled to an extension of time under 2·19·2. Clause 3·18·4 refers to a Code of Practice, which is included in the form in Schedule 4. Its purpose is to 'help in the fair and reasonable operation' of the provisions regarding further testing. It sets out criteria that the contract administrator should consider when deciding whether to instruct further testing, including, for example, the potential consequences of the non-compliance and the standard of supervision of the work by the contractor.

Non-compliance with clause 2·1

6.50   Clause 2·1 requires the contractor to carry out the work 'in a proper and workmanlike manner' and in accordance with the Health and Safety Plan. Clause 3·19 states that in the event of any failure to comply in this respect, the contract administrator may issue instructions requiring compliance, and these will not result in any addition to the contract sum, nor will they entitle the contractor to any extension of time or direct loss and/or expense. The clause empowers the contract administrator to intervene in the contractor's working methods if necessary.

Sub-contracted work

6.51   SBC05 provides for two methods of sub-contracting work; to sub-contractors selected by the contractor or to those 'listed' in the contract documents by the employer. Both of these methods allow for some control over which firms the contractor uses.

Contractor selected sub-contractors

6.52   Under clause 3·7 the contractor may only sub-contract work, including the design of the CDP, with the written consent of the contract administrator. Failure to obtain this would be a default, providing grounds for termination under clause 8·4·1·4. Clause 3·7 states, however, that the contract administrator's permission cannot be unreasonably withheld. It is suggested that permission is required for each instance of sub-letting, rather than agreeing to sub-letting in principle. There is no requirement to use a particular form of sub-contract although JCT Ltd now publish a suite of sub-contracts developed for use with SBC05, including one for use where the sub-contracted work relates to the Contractor's Designed Portion. Whatever form of domestic sub-contract is used, clause 3·9 requires that it must include certain conditions, including that:

- the sub-contract is terminated immediately upon termination of the main contract (cl 3·9·1);

- unfixed materials and goods placed on the site by the sub-contractor shall not be removed without written consent by the contractor (cl 3·9·2·1);

- it shall be accepted that materials or goods included in an Interim Certificate that have been paid by the employer become the property of the employer (cl 3·9·2·1·1);

- it shall be accepted that any materials or goods paid for by the main contractor prior to being included in a certificate become the property of the main contractor (cl 3·9·2·1·2);

- the sub-contractor shall provide access for the contract administrator to workshops etc. (cl 3·9·2·2);

- the sub-contractor has a right to interest on late payments by the contractor at the same rate as that due on main contract payments (cl 3·9·2·3);

- the sub-contractor will enter into warranties as required under the main contract (cl 3·9·2·4);

- neither clause 3·9·2·1·1 or 3·9·2·1·2 shall affect the vesting in the contractor of property in any listed item (cl 3·9·2·5).

6.53   Clause 3·9·2·1 is to protect the position of the employer and the provisions regarding unfixed goods, and materials are of particular importance in this respect. If a main contractor should sub-contract on other terms, and this results in losses to the employer, then the contractor may be liable as this would be a breach of contract.

6.54   Once materials have been built in, under common law they would normally become the property of the owner of the land, irrespective of whether or not they have been paid for by the contractor. This would be the case even if there were a retention of title clause in the contract with the sub-contractor or supplier. A retention of title clause is one which stipulates that the goods sold do not become the property of the purchaser until they have been paid for, even if they are in the possession of the purchaser.

6.55   The employer could be at risk, however, where materials have not yet been built in, even where the materials have been certified and paid for. The contractor might not actually own the materials paid for because of a retention of title clause in the sale of materials contract. Under the Sale of Goods Acts 1979, sections 16–19, property in goods normally passes when the purchaser has possession of them, but a retention of title clause will be effective between a supplier and a contractor even where the contractor has been paid for the goods, provided they have not yet been built in. It should be noted, however, that the employer may have some protection through the operation of section 25 of the Act, which in some circumstances allows the employer to treat the contractor as having authority to transfer the title in the goods, even though this may not in fact be the case (see *Archivent* v *Strathclyde Regional Council* in paragraph 6.58).

6.56   Another risk relating to rightful ownership is where the contractor fails to pay a domestic sub-contractor who has purchased materials, and the sub-contractor claims ownership of the unfixed materials. Here the risk may be higher, as a work and materials contract is not governed by the Sale of Goods Act. Therefore there can be no assumption that property would pass on possession.

6.57   SBC05 attempts to deal with the issues surrounding ownership in several ways. First, unfixed materials and goods, which have been delivered to the site and intended for the Works, may not be removed without the written consent of the contract administrator (cl 2·24 and 3·9·3). Removal would be a breach of contract, therefore the employer could claim from the contractor for any losses suffered through unauthorised removal. This would apply even though the materials or goods may not yet have been included in any certificate. Secondly, unfixed materials and goods either on or off site which have been included in a certificate which has been paid are to become the property of the employer (cl 2·24 and 2·25), and the contractor is thereby prevented from disputing ownership.

6.58   Clauses 2·24 and 2·25 of the main contract, however, are only binding between the parties, and do not place obligations on any sub-contractor. The risk facing the employer is that if the contractor becomes insolvent, a sub-contractor or supplier may still have a rightful claim to ownership of the unfixed goods, even though they have been paid for by the employer (see *Dawber Williams Roofing* v *Humberside County Council* below). The main contract therefore requires that all sub-contracts include similar clauses to 2·24 and 2·25 regarding non-removal from site, and ownership passing upon payment (cl 3·9·2·1). Sub-contracts must also include a clause stating that once materials and goods have been certified and paid for under the main contract they become the property of the employer and that the sub-contractor 'shall not deny' this. This would operate even where the main contractor has become insolvent. Even this might not protect the employer in some circumstances because if the sub-contractor does not have 'good title' it cannot pass it on. Thus, for example, it might not prevent a sub-sub-contractor claiming ownership.

Archivent Sales & Developments Ltd v Strathclyde Regional Council (1984) 27 BLR 98
(Court of Session, Outer House)

Archivent agreed to sell a number of ventilators to a contractor who was building a primary school for Strathclyde Regional Council. The contract of sale included the term 'Until payment of the price in full is received by the company, the property in the goods supplied by the company shall not pass to the customer.' The ventilators were delivered and included in a certificate issued under the main contract (JCT63), which was paid. The contractor went into receivership before paying Archivent, who claimed against the Council for the return of the ventilators or a sum representing their value. The Council claimed that section 25(1) of the Sale of Goods Act operated to give them an unimpeachable title.

The judge found for the Council. Even though the clause in the sub-contract successfully retained the title for the sub-contractor, the employer was entitled to the benefit of section 25(1) of the Sale of Goods Act. The contractor was in possession of the ventilators and had ostensible authority to pass the title on to the employer, who had purchased them in good faith.

Dawber Williams Roofing Ltd v Humberside County Council (1979) 14 BLR 70

The plaintiffs entered into a sub-contract with Taylor and Coulbeck Ltd (T&C) to supply and fix roofing slates. The main contractor's contract with the defendant was on JCT63. By clause 1 of their sub-contract (which was on DOM/1) the plaintiffs were deemed to have notice of all the provisions of the main contract, but it contained no other provisions as to when property was to pass. The plaintiffs delivered 16 tons of roofing slates to the site, which were included in an Interim Certificate, which was paid by the defendant. T&C then went into liquidation without paying the sub-contractor, who brought a claim for the amount or alternatively the return of the slates. The judge allowed the claim, holding that clause 14 of JCT63 could only transfer property where the main contractor had a good title. (The difference between this and the *Archivent* case above is that in this case the sub-contract was a contract for work and materials, to which the Sale of Goods Act did not apply.) Provisions within clause 3·9·2 of SBC05 now deal with the problem illustrated by this case.

Listed sub-contractors

6.59   Under clause 3·8 the contractor's choice of a sub-contractor to carry out certain work measured or otherwise described can be restricted to any one of three or more persons named in the contract bills, or in a list annexed to the bills. The contractor must select one of those listed (cl 3·8·1). The contractor is responsible for the performance of such sub-contractors to the same degree as it would be for any sub-contractor it had selected itself. All sub-contracts should contain the provisions described above and the contractor may be required to obtain warranties from the listed sub-contractor (see paragraphs 2.19 and 3.57).

6.60   Names may be added to the list by either the employer or the contractor, with the consent of the other. If less than three of those listed are able or willing to carry out the work then the employer must add names to bring the total up to no fewer than three (cl 3·8·3 see Figure 19). If it is not possible to maintain a list of three, then the work may be carried out by the contractor, or sub-let to a domestic sub-contractor under clause 3·7. The contractor has the right of reasonable objection to any new addition to the list, and given that it is taking responsibility for the subcontractor's performance, it is likely that any real concern about the competence of the firm would be a good reason for objecting. In respect of any suggestions by the contractor, the form again states that the employer's consent 'shall not be unreasonably delayed or withheld' (cl 3·8·2). Where there are difficulties in engaging any of those preferred by the employer, the employer may be considered unreasonable if it objects to any alternative suggestions put forward by the contractor.

Work not forming part of the contract/persons engaged by the employer

6.61   Under clause 2·7 the employer may engage persons direct to carry out work that does not form part of the contract, while the main contractor is still in possession. If the contract bills

Figure 19 Appointment of listed subcontractors

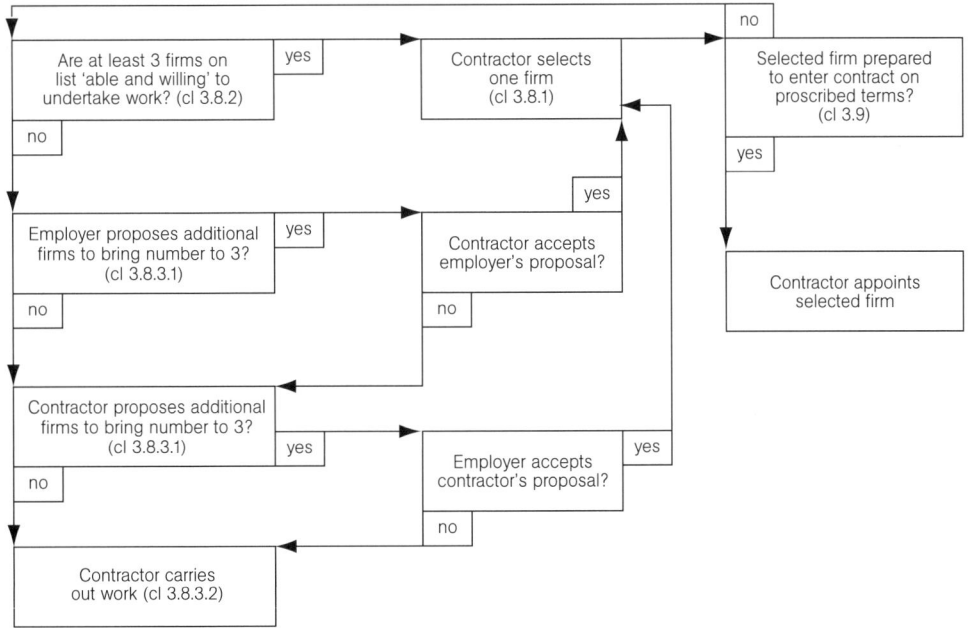

have included this requirement, then the contractor must permit the employer to execute such work (cl 2·7·1). Otherwise the employer can only do this with the contractor's permission (cl 2·7·2). The permission may not be unreasonably delayed or withheld.

6.62   It should be remembered that the indemnities given by the contractor under clause 6·1 (and therefore the insurances under clause 6·4) do not extend to persons employed under clause 2·7. The employer should also be made aware that any serious disruption caused to the contractor's working could lead to an extension of time (cl 2·29·6), to loss and expense (cl 4·24·5), or even to termination (cl 8·9·2).

Making good defects

6.63   The contract administrator may instruct the contractor to make good any 'defects, shrinkages or other faults' which appear during the rectification period (cl 2·38·2). The power is limited to those defects that result from the work not having been carried out in accordance with the contract or to the contractor failing to comply with its obligations with respect to the designed portion. This does not include other defects that may be due, for example, to errors in the design information supplied to the contractor, or to general wear and tear resulting from occupation by the employer. The power is also limited to those defects that appear after practical completion, although it would be sensible to allow the contractor to correct any outstanding defects (*Perce and High* v *Baxter*). The power would cover defects appearing during the rectification period that have been caused by frost, but only where the damage was due to a default of the contractor before practical completion.

6.64   If an instruction to make good defects is issued, the contractor must comply 'within a reasonable time' (cl 2·38). Unless causing unacceptable problems, defects that appear during the rectification period are normally left until after the expiration of the period. The power to instruct ends at the time the defects schedule is issued or 14 days after the end of the rectification period.

6.65   The contract administrator has a duty under clause 2·38·1 to issue a schedule of defects not later than 14 days after the end of the defects liability period, the only point where the contract requires the contract administrator to issue such a schedule. The schedule is issued in the form of an instruction, and the contractor is required to rectify the defects within a reasonable time. If the contract administrator, with the agreement of the employer, decides to accept the defective work then this should be clearly shown in the instruction, and under clause 2·38 an appropriate deduction is made from the contract sum. Care should be taken to establish the full extent of the problem before such a course of action is taken, as it is unlikely that the employer would thereafter be able to claim for consequential problems or further remedial work. In particular, if the work was stated to be 'to approval' in the contract documents, this will have implications for the Final Certificate (see paragraphs 8.39–8.41) which should be resolved before the work is accepted.

6.66   Once satisfied that all the defects have been made good, the contract administrator must issue a certificate to that effect (cl 2·39). The certificate is one of the pre-conditions to the issue of the Final Certificate. The contract does not state what should happen in respect of defects which appear after the issue of the certificate but before the issue of the Final Certificate. It is, however, clear from clause 2·38 that the contract administrator no longer has the power to instruct that these are made good. It is suggested that in such circumstances there would be two possible courses of action. The first would be to make an agreement with the contractor to rectify the defects before the Final Certificate is issued. If the contractor refused to do this an amount could be deducted from the contract sum to cover the cost of making good the work, but this would involve some risk to the employer. The second and less risky course would be to have the defective work rectified by another contractor, and deduct the amount paid from the contract sum. This would involve a delay to the issue of the Final Certificate and would probably be disputed by the contractor.

Pearce and High v John P. Baxter and Mrs A. S. Baxter [1999] BLR 101 (CA)

The Baxters employed Pearce and High on MW80 to carry out certain works at their home in Farringdon. Following practical completion the contract administrator issued Interim Certificate number 5, which the employer did not pay. The contractor commenced proceedings in Oxford County Court, claiming payment of that certificate and additional sums. The employer in its defence and counterclaim relied on various defects in the work that had been carried out. Although the Defects Liability Period had by that time expired, neither the contract administrator nor the employers had notified the contractor of the defects. The Recorder held that clause 2·5 was a condition precedent to the recovery of damages by the employer, and further stated that it was a condition precedent that the building owner has notified the contractor of patent defects within the Defects Liability Period.

The employer appealed and the appeal was allowed. Lord Justice Evans stated that there were no clear express provisions within the contract which prevented the employer bringing a claim for defective work, regardless of whether notification had been given. He went on to state, however, that the contractor would not be liable for the full cost to the employer of remedying the defects, if the contractor had been effectively denied the right to return and remedy the defects itself.

7   Sums properly due

7.1   SBC05 With Quantities and Without Quantities are both lump sum contracts. The contract sum will be the tender figure accepted or agreed following negotiation and is entered in article 2. However, this is rarely the amount actually paid. The wording of the contract recognises this by the qualifying reference 'or such other sum as shall become payable' (article 2).

7.2   The contract sum may contain provisional sums or approximate quantities to cover the cost of work that cannot be accurately described or measured until work is under way. Most contracts will require some variations to the Works. There is also the possibility of claims from the contractor for loss and expense arising from intervening events that could not be foreseen at the time of tendering. While it is possible in theory to make a contract 'fixed price', most will allow for fluctuations to some degree. Fees or charges in respect of statutory matters which are not allowed for in the contract bills will require an adjustment to the sum. VAT is, of course, not included in the contract sum.

7.3   There will therefore almost inevitably be adjustments to the contract sum; SBC05 clause 4·2, however, makes it clear that the only alterations that may be made are those provided for in the terms. Items to be included in adjustments are set out in clause 4·3, and the ascertained amounts will be added or deducted as appropriate at the periods for certification (cl 4·4).

7.4   Arithmetical errors by the contractor in pricing are not allowed as a cause for adjustment. Errors in the preparation of the contract bills, on the other hand, must be corrected and will then be treated as if they are a variation (cl 2·14). Any divergence between the contract drawings and other documents, which necessitates an instruction by the contract administrator may also result in a variation (cl 2·15).

7.5   SBC05 With Approximate Quantities is a re-measurement contract. Approximate quantities only are given for all of the work, and the contractor submits a fully priced copy of the bills of approximate quantities at tender stage, which forms the basis of the contract. No 'contract sum' is entered in the Articles. All the work is re-measured prior to certification and the contractor is paid for the actual quantities of work carried out. The final amount payable in accordance with the conditions is termed the 'Ascertained Final Sum' (Article 2).

An approximate quantity

7.6   Under SBC05 With Quantities, where work can be described in accordance with the Standard Method of Measurement, but where the quantity involved is uncertain, an 'Approximate Quantity' can be included in the bills. The contract administrator is not required to issue any further instruction for the contractor to carry out this work. After it has been carried out, the work is valued as described in clause 5·6·1 (discussed at paragraph 6.17).

7.7   Difficulties can arise if the approximate quantity is not a reasonably accurate forecast of the quantity of work required. Then the valuation must include a fair allowance for the difference in quantity over and above the rates or prices tendered by the contractor (cl 5·6·1·4). The contractor may also claim that the inaccuracy is a relevant event under clause 2·29·4 and a matter for direct loss or expense under clause 4·24·4. Similar valuation rules are set out in the With Approximate Quantities version of SBC05.

Provisional sums

7.8   If insufficient information can be provided at the time of tender to allow an item to be described and measured in accordance with the Standard Method of Measurement, then a provisional sum may be inserted in the bills to cover the item. The contract administrator must issue instructions regarding all work covered by provisional sums in the contract bills and the contractor can take no action with regard to this work until receiving an instruction (cl 3·16). Under the Standard Method of Measurement, provisional sums are either for defined work or for undefined work. Provisional sums may also be included in the specification/schedules of work in the Without Quantities version. These are dealt with in the same way as those for undefined work described below. In all cases the work covered by a provisional sum is valued by the quantity surveyor as described below.

Defined work

7.9   The information required to place provisional work in the defined category is listed in SMM7 General Rule 10·3. The tenderer must be aware of the nature and construction of the work, how and where the work fits into the building, the scope and extent of the work and any specific limitations on method or sequence or timing. In other words, the description must be sufficiently detailed for the contractor to make proper allowance for the effect of the work when pricing the relevant preliminaries, and to allow for the work in the programme.

7.10   If the information provided is not as detailed as the Rule requires, or if it is erroneous, then a corrective instruction is required from the contract administrator (cl 2·14·1). This will be treated as a variation (cl 2·14·3) and could give rise to a notice of delay (cl 2·29·1) and an application for reimbursement of direct loss and expense (cl 4·24·1) from the contractor. The corrective action cannot simply be to change from the defined category to the undefined category by substituting a new provisional sum.

Undefined work

7.11   A provisional sum for undefined work will be applicable where it is not possible to supply the amount of information needed to comply with Rule 10·3. The contractor will not have been able to make proper allowance for the work in programming, planning, or the pricing of preliminaries. A provisional sum in this case should be sufficient to not only cover the net cost, but also to take into account the fact that there might be additions to preliminaries, attendance, and perhaps loss and expense, etc. There is also a risk that the contractor might give notice of delay arising from the contract administrator's instruction. Unlike an instruction for the expenditure of a provisional sum for defined work, that for a provisional sum for undefined work could be a Relevant Event (cl 2·29·2·1) and a 'matter' for which a loss and expense application can be made (cl 4·24·2·1).

7.12   Provisional sums may be included for items that are not specifically work, for example testing, site boards, site facilities, etc. The heading in the bills of quantities will simply be 'include the Provisional Sum of _____ for _____', or some other appropriate wording. For work to be carried out by statutory authorities it is suggested that the description of the work be followed by a similar heading.

Valuation of variations

7.13   There are two mechanisms by which a variation can be valued under the provisions of the contract: (1) acceptance of a Schedule 2 quotation; (2) valuation by the quantity surveyor (see Figure 20). Of course, as indicated in clause 5·3·1, it is always open for the contractor and employer to simply agree a price.

Schedule 2 quotations

7.14   If the employer or the contract administrator wishes to ascertain the contractor's price for a variation, then the instruction requiring the variation should request that a quotation is submitted in accordance with Schedule 2 (cl 5·3·1). The contractor has seven days to object to the application of this procedure. If the contractor objects, the instruction is not carried out unless the contract administrator instructs that it should be, in which case it is valued 'by a Valuation', i.e. by the quantity surveyor (cl 5·3·2). The instruction should include sufficient detail to enable the contractor to provide the information required. Any addendum to the contract bills issued for the purposes of obtaining a Schedule 2 quotation should be prepared according to SMM7 unless otherwise stated (cl 2·13·1). Any errors in the addendum are to be treated in the same way as errors in the contract bills (cl 2·14·1).

7.15   If no objection is raised, the contractor must submit the Schedule 2 quotation within 21 days of receipt of the instruction or any additional information requested under Schedule 2:1·1. The quotation should identify the direct cost of complying with the instruction, the period required for extension to the contract period and the sum acceptable in lieu of direct loss and/or expense. The quotation should make reference where relevant to rates and prices in the contract bills (Schedule 2:2·1).

7.16   If accepted, the quotation takes the place of valuation by the quantity surveyor. This method brings certainty of outcome for the parties, as both are bound by what is agreed with respect to the value of the work, the extension of time and the direct loss and/or expense. In addition, the fluctuations provisions do not apply to the accepted quotation (cl 4·22). The certainty, however, is likely to be secured only at a price, particularly where the variation does not relate to work for which there are rates and prices in the bills. If the quotation is rejected, the work can still be instructed but is then subject to valuation by the quantity surveyor (Schedule 2:4·1). The contractor is paid a fair and reasonable amount for the cost of preparing the Schedule 2 quotation (Schedule 2:5).

Figure 20 Valuation of variations

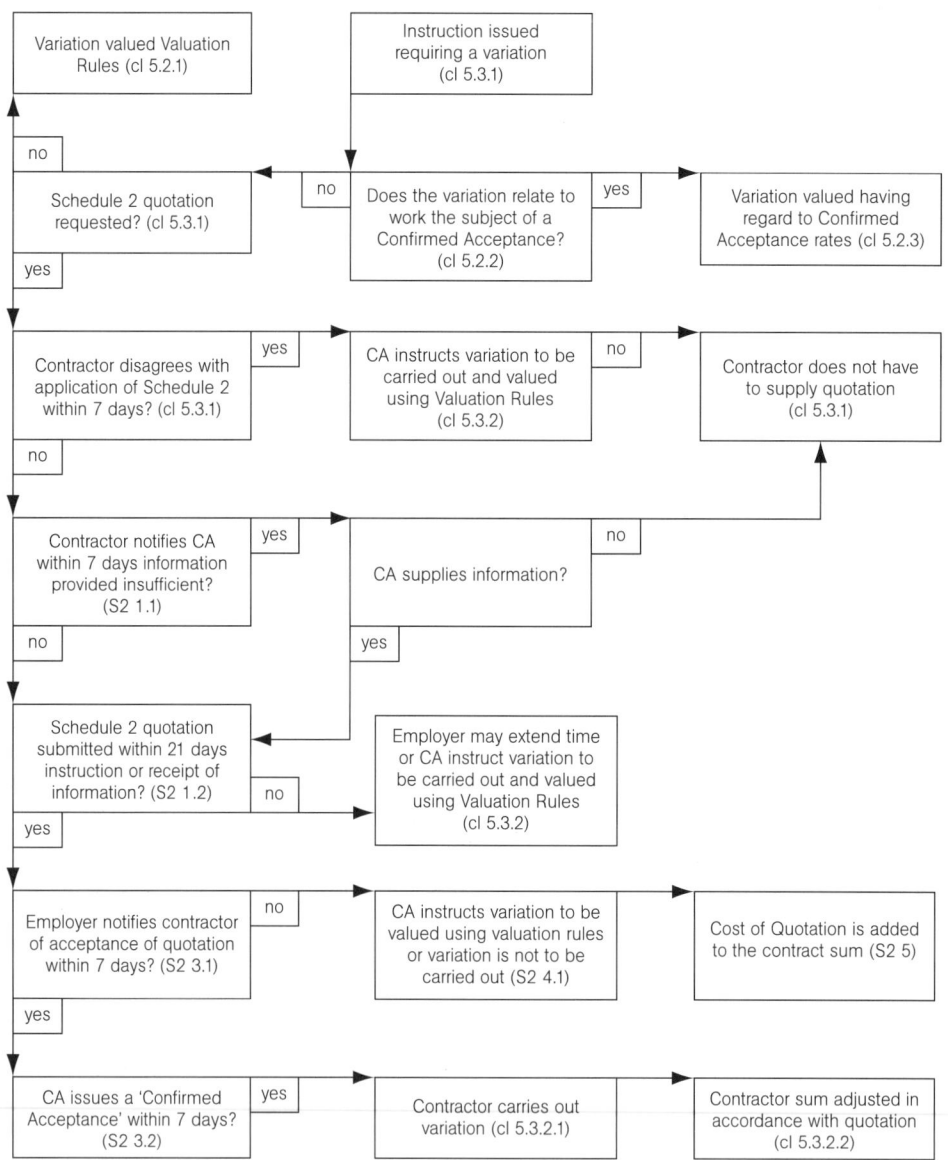

7.17 If the contract administrator subsequently issues a variation to work for which a Schedule 2 quotation has been given and accepted, then this variation is valued on a fair and reasonable basis by the quantity surveyor, 'having regard to' the contents of the original Schedule 2 quotation (cl 5·3·3). This provision appears to exclude possible re-assessment by the quantity surveyor. In other words, the parties will be bound by the terms agreed in the original Schedule 2 quotation with respect to both the original instruction and to any future related variations.

7.18   It should be noted that the Schedule 2 provisions do not appear to apply to instructions regarding the expenditure of provisional sums, as clause 5·3·1 refers only to 'Variations'. However, there would be nothing to prevent the contract administrator or employer requesting a quotation prior to issuing such an instruction.

Valuation by the quantity surveyor

7.19   If no Schedule 2 quotation is sought, or if the quotation is rejected, then the valuation of the variation must be made by the quantity surveyor according to the rules set out in clause 5·6·1 (cl 5·2·1). If the variation involves an omission then the value of the work shown in the contract bills is deducted from the contract sum (cl 5·6·2).

7.20   Clause 5·6·1 (With Quantities version) includes for:

- work of similar character undertaken under similar conditions and where the quantity does not change significantly (cl 5·6·1·1);

- work of similar character but not undertaken under similar conditions and/or where the quantity changes significantly (cl 5·6·1·2);

- work not of similar character (cl 5·6·1·3).

7.21   In the first two cases, bills of quantities' rates and prices are to be used in assessing the value of the variation, and it should be noted that the work is not necessarily identical, and that the contractual rates must be used even where those figures contain errors (*Henry Boot Construction Ltd* v *Alstom Combined Cycles*). In the third case the work should be valued at 'fair rates and prices'. Dissimilar conditions might include, for example, that the instructed work is carried out in winter, whereas under the bills it had been assumed it would be carried out in summer. Such an assumption, however, would have to be clear from an objective analysis of the contract documents (*Wates Construction* v *Bredero Fleet*).

Henry Boot Construction Ltd v Alstom Combined Cycles [2000] BLR 247

By a contract formed in 1994 ACC employed Henry Boot to carry out civil engineering works in connection with a combined cycle gas turbine power station for PowerGen plc at Connah's Quay in Clwd. During post tender negotiations, a price of £258,850 was agreed for temporary sheet piling to trench excavations. Disputes arose regarding the valuation of this work, which were initially taken to arbitration. The arbitrator found that the agreed figure contained errors that effectively benefited Boot. Boot argued that nevertheless the figure should be used to value the work under clause 52(1). The arbitrator decided that 52(1) a and b were inapplicable, and that 52(2) should be applied to achieve a fair valuation. Boot appealed to the TCC, and Judge Humphrey Lloyd decided that the mistake made no difference; the agreed rate should be used even if the results were unreasonable. Clause 52(2) created only a limited exception where the scale or nature of the variation itself made it unreasonable to use the contract rates.

Wates Construction (South) Ltd v Bredero Fleet Ltd (1993) 63 BLR 128

Wates Construction entered into a contract on JCT80 to build a shopping centre for Bredero. Some sub-structural work differed from that shown on the drawings and disputes arose regarding the valuation of the Works, which were taken to arbitration. In establishing the conditions which, under the contract, it had been assumed the work would be carried out, the arbitrator took into account pre-tender negotiations and the actual knowledge that Wates gained as a result of the negotiations, including proposals that had been put forward at that time. Wates appealed and the court found that the arbitrator had erred by taking this extrinsic information into consideration. The conditions under which the Works had to be executed had to be derived from the express provisions of the bills, drawings and other contract documents.

7.22   Clause 5·7 (Daywork) provides for 'work which cannot properly be valued by measurement'. In such cases the valuation is based on the prime cost of the work, calculated in accordance with the definitions of prime cost referred to in 5·7·1 and 5·7·2. The contractor must provide vouchers for verification by the contract administrator, showing specified details of the daywork, by the end of the week following that in which the work is carried out.

7.23   Where, as a result of a variation, other contract work has to be carried out under different conditions, then this must be treated as if it were a variation and valued accordingly, even though the consequences were not themselves identified in the original instruction (cl 5·9).

7.24   In all cases any measurement should be made according to the principles governing the preparation of the contract bills (i.e. SMM7, unless otherwise stated therein, see clauses 2·13·1 and 5·6·3). Clause 5·6·3·3 authorises appropriate allowance to be made for an addition to or reduction of preliminaries, except for instructions regarding the expenditure of provisional sums for defined work.

7.25   In the Without Quantities version the clause 13·5·1 valuation rules are slightly simpler. Work of a similar character is to be valued according to rates and prices in the Priced Document, with a fair allowance being made if there is any change in conditions under which the work is carried out, or any significant change in quantity. Where the work is not of similar character it should be valued at fair rates and prices.

7.26   The value of variations to the Contractor's Designed Portion 'shall be consistent with the values of work of a similar character set out in the CDP Analysis, making due allowance for any change in the conditions under which the work is carried out... Where there is no work of a similar character... a fair valuation shall be made' (cl 5·8·2). The 5·6·3, 5·7 and 5·9 rules apply as relevant.

Reimbursement of direct loss and/or expense

7.27   The objective of the 'Loss and Expense' provisions is to enable the contractor to be reimbursed for direct loss and/or direct expense suffered as a result of delay or disruption, and for which the contractor is not reimbursed under any other provision in the contract. As an alternative the contractor may be able to pursue this as a claim for general damages for breach of contract at common law (cl 4·26).

7.28   The contractor must apply in writing, otherwise the contract administrator has no power or obligation to deal with loss and expense (cl 4·23). The application must be submitted promptly, in fact the use of the phrase 'is likely to occur' suggests an anticipatory quality. No particular format is specified but the application should be supported by such detail as is reasonably necessary, including identifying all 'matters' concerned (*London Borough of Merton* v *Leach*). The contract administrator or quantity surveyor can request additional information.

London Borough of Merton v Stanley Hugh Leach Ltd (1985) 32 BLR 51 (Chancery Division)

Stanley Hugh Leach entered into a contract with the Borough to construct 287 dwellings. The contract was JCT63. There was a considerable delay on the contract and a dispute arose regarding this delay and related claims for loss and expense. The dispute went to arbitration and the arbitrator made an interim award on a list of matters. Merton appealed and the court considered 15 questions framed as preliminary issues. Among other things the court stated that applications for direct loss and/or expense must be made in sufficient detail to enable the contract administrator to form an opinion as to whether there is any loss and/or expense to be ascertained. If there is, then it is the responsibility of the contract administrator to obtain enough information to reach a decision. This could of course include requesting information from the contractor. The court also held that the application must be made within a reasonable time and not so late that the contract administrator can no longer form an opinion on matters relevant to the application.

7.29   Claims can only be made for loss and expense suffered through deferment of possession, or the particular Relevant Matters listed in clause 4·24. Other losses are irrecoverable under the contract, although disputed claims may be referred to adjudication, arbitration or litigation. The matters listed in clause 4·24 are concerned with situations where the loss or expense is attributable to the employer, and excludes the 'neutral causes' which feature in the extension of time provisions of clause 2·29. Clause 4·23 includes for two types of claim, often difficult to separate because they are perhaps based on the same facts. The phrase 'regular progress of the Works or of any part thereof has been or is likely to be so materially affected' allows for disruption not anticipated when tendering, and prolongation for which an extension of time may have been awarded. Any disruption should be related to the progress necessary to complete by the completion date, not necessarily the actual sequences of events on site.

7.30   In ascertaining the loss and expense, the contract administrator must determine what has actually been suffered. The sums that can be awarded can include any loss or expense that has arisen directly as the result of the Relevant Matter. The loss and expense award is in effect an award of damages, and the contract administrator should approach its assessment on the same principles as a court would in awarding damages for breach of contract. In broad terms, the object is to put the contractor back into the position in which it would have been but for the disturbance. The contractor ought to be able to show that it has taken reasonable steps to mitigate its loss.

7.31   The following are items which could be included:

- increased preliminaries;

- overheads;

- loss of profit;

- uneconomic working;

- increases due to inflation; and

- interest or finance charges.

7.32   The items claimed must be things which the contractor could not recover under any other term of the contract (for example, it must not duplicate a claim under clause 5·6). Prolongation costs such as on-site overheads would normally only be claimable for periods following the date for completion. (For head office overheads, etc., see *McAlpine* v *Property and Land Contractors* below.) Interest may also be recoverable, but only if it can be proved to have been a genuine loss (*F. G. Minter* v *WHTSO*). As clause 4·23 refers to losses which are 'likely to occur', the award need not be restricted to suffered prior to the time the contractor's application is made, but could include those suffered up to the date of the ascertainment, (this would apply particularly to financing charges), and could arguably be extended to losses that could be predicted as likely to occur up to the date the reimbursement is made.

F. G. Minter Ltd v Welsh Health Technical Services Organisation (1980) 13 BLR 1 (CA)

Minter was employed by Welsh Health Technical Services Organisation (WHTSO) under JCT63 to construct the University Hospital of Wales (second phase) Teaching Hospital. During the course of the contract several variations were made, and the progress of the Works was affected due to the lack of necessary drawings and information. The contractor was paid amounts in respect of direct loss and/or expense, but the amounts paid were challenged as insufficient. The amounts had not been certified and paid until long after the losses had been incurred, therefore the amounts should have included an allowance in respect of finance charges or interest. Following arbitration several questions were put to the High Court, including whether Minter was entitled to finance charges in respect of any of the following periods:

(a)   between the loss and/or expense being incurred and the making of a written application for the same;

(b)   during the ascertainment of the amount; and/or

(c)   between the time of such ascertainment and the issue of the certificate including the ascertained amount.

The court answered 'no' to all three questions and Minter appealed. The Court of Appeal decided that the answer was 'yes' to the first question and 'no' to the other two.

Alfred McAlpine Homes North Ltd v Property and Land Contractors Ltd (1995) 76 BLR 59

An appeal arose on a question of law arising out of an arbitrator's award regarding the basis for awarding direct loss and expense with respect to additional overheads and hire of small plant, following an instruction to postpone the Works. The judgment contains useful guidance on the basis for awarding direct loss and expense. To 'ascertain' means to 'find out for certain'. It is not necessary to differentiate whether a head of claim is 'loss' or 'expense'. Regarding overheads, a contractor would normally be entitled to recover as a 'loss' the shortfall in the contribution that the volume of work had been expected to make to the fixed head office overheads, but which, because of a reduction in volume and revenue caused by the prolongation, was not in fact made. The fact that Emden or Hudson formulae depend on certain assumptions mean that they are frequently inappropriate. The losses on the plant should be the true cost to the contractor, not based on notional or assumed hire charges.

7.33   The contractor must provide full details and particulars of all items concerned with the alleged loss or expense. These should identify which of the losses claimed relate to each of the 'matters' that have occurred. This is sometimes compromised by the use of a 'rolled up' or composite claim approach where it is not really practicable to separate and itemise the effect of a number of causes. This has been accepted by the courts provided that as much detail as possible has been given, and provided that all disturbance was due to matters under clause 26, and not caused by the contractor.

7.34   Formulae such as the Hudson or Emden formulae are sometimes used for estimating head office overheads and profit, which may be difficult to substantiate. These can only be used where it has been established that there has been a loss of this nature. To do this the contractor must be able to show that, but for the delay, the contractor would have been able to earn the amounts claimed on another contract, for example by producing evidence such as invitations to tender which were declined. Such formulae may be useful where it is difficult to quantify the amount of the alleged loss, provided a check is made that the assumptions on which the formula is based apply.

7.35   Although direct loss and/or expense is a matter of money, not time, which are quite separate issues, there is often a practical correlation in the case of prolongation. Any general implication, however, that there is a link would be incorrect and in principle disruption claims and delay to progress are independent. An extension of time, for example, is not a condition precedent to the award of direct loss and/or expense (*H. Fairweather & Co.* v *Wandsworth*).

7.36   Though the contract administrator may delegate the duty of ascertaining the direct loss and expense, it does not appear that it is obligatory to accept the quantity surveyor's opinion (*R. Burden* v *Swansea Corporation*), although the quantity surveyor's assessment would be strong evidence as to what the correct amount should be. Applications or claims from the contractor or nominated sub-contractor made under the contract must be dealt with according to the procedures of the contract. Failure to certify an amount properly due will not prevent recovery, and could leave the employer liable in damages for breach of contract (*Croudace* v *London Borough of Lambeth*).

R. Burden Ltd v Swansea Corporation [1957] 3 All ER 243 (HL)

Burden entered into a contract with Swansea Corporation to build a school. The contract provided for Interim Certificates to be issued at intervals by the contract administrator. The contract administrator, who was the Corporation's borough contract administrator, acted originally as both contract administrator and surveyor under the contract. Later, after 20 certificates had been issued, the firm of quantity surveyors who had originally prepared the bills were appointed to act as surveyors under the contract in place of the borough contract administrator. In the next certificate the surveyor reduced the amount applied for by the contractor by around 75 per cent, and the contract administrator certified the lower figure. The surveyor later discovered that he had made a mistake, but did not inform the contract administrator of the error. The contractor gave notice determining the contract, on the grounds that the employer had interfered with the issue of the certificate. The House of Lords decided that a mistake in a direction as to the amount to be paid did not amount to interference or obstruction. It was suggested that the contract administrator would have been at liberty to have certified a different amount if aware of the error.

Croudace Ltd v The London Borough of Lambeth (1986) 33 BLR 25 (CA)

Croudace entered into an agreement with the London Borough of Lambeth to erect 148 dwelling houses, some shops and a hall. The contract was on JCT63 and the contract administrator was the borough's chief contract administrator and the quantity surveyor was the borough's chief quantity surveyor. The contract administrator delegated his duties to a private firm of contract administrators. Croudace alleged that there had been delays and that they had suffered direct loss and/or expense, and sent letters detailing the matters to the contract administrators. In reply, the contract administrators told Croudace that they had been instructed by Lambeth that all payments relating to 'loss and expense' had to be approved by the Borough. The chief contract administrator of Lambeth then retired and was not immediately replaced. There were considerable delays pending a further appointment and Croudace began legal proceedings. The High Court found that the Borough were in breach of contract in failing to take the necessary steps to ensure that the claim was dealt with, and were liable to Croudace for this breach. The Court of Appeal upheld this finding.

Fluctuations

7.37   Depending upon the economic climate it may be to the employer's advantage to ask for literally a 'fixed' or 'guaranteed' price. However, the term 'fixed price' in building contracts usually includes for limited fluctuations, for example tax changes.

7.38   SBC05 allows for 'Tax etc.' fluctuations (Option A); the traditional full fluctuations in labour and materials (Option B); and the use of price adjustment formulae (Option C). Clause 4·21 sets out the three options and, if not identified in the Contract Particulars, then Option A the 'Tax etc.' (i.e. so-called fixed price) applies. The fluctuations clauses are included under Schedule 7.

7.39   Option A provides for full recovery of all fluctuations in the rates of contributions, levies and taxes in the employment of labour, and in the rates of duties and taxes on the procurement of materials. In short, the only amounts payable are those arising out of an Act of Parliament or delegated legislation. Option B allows, in addition, for fluctuations in the actual market costs of labour and materials. Contractors, however, have pointed out that many less obvious increases are nevertheless not included; therefore a 'percentage addition' is made to allow for these. The agreed percentage is entered in the Contract Particulars. Option C allows for adjustment based on the use of formulae: it does not necessarily take account of the actual costs but is relatively simple to operate and is generally considered by contractors to be a fair adjustment.

7.40   Where a contract includes for fluctuations, in the absence of anything to the contrary, they will be payable for the whole time the contractor is on site even though it fails to complete within the contract period (*Peak (Liverpool) Ltd* v *McKinney Foundations Ltd*). There is a so-called 'freezing' provision in SBC05 clause Option A·4·7 (and corresponding clauses), but this depends on the clauses relating to extensions of time being left unamended, and all notices of delay properly dealt with by the contract administrator.

7.41   Where the price of a variation has been agreed following acceptance of a Schedule 2 quotation, the fluctuation provisions do not apply to that item (cl 4·22).

Peak Construction (Liverpool) Ltd v McKinney Foundations Ltd (1970) 1 BLR 111 (CA)

Peak Construction were main contractors on a contract to construct a multi-storey block of flats for Liverpool Corporation. As a result of defective work by nominated sub-contractors, McKinney Foundations' work on the main contract was halted for 58 weeks, and the main contractors brought a claim against the sub-contractors for damages. The Official Referee at first instance found that the entire 58 weeks delay was caused by the nominated sub-contractor, and awarded £40,000 of damages, £10,000 of which was for rises in wage rates during the period. McKinney appealed, and the Court of Appeal found that the award of £10,000 could not be upheld as clause 27 of the main contract entitled Peak Construction to claim this from the Corporation right up until the time when the work was halted.

---

8 Payment

8.1 Certification is one of the most important duties of the contract administrator under the contract, and one to be carried out with care. It is not unheard of for contractors to become insolvent during the course of a contract, and if the certificates have been overvalued the employer may suffer losses which could have been avoided. On the other hand, the contractor has a right to be paid what the terms of the contract state is due, and the contract administrator must not be influenced by any attempt on the part of the employer to delay certification or withhold amounts properly due.

8.2 Failure to certify correctly could amount to negligence. In the case of *Sutcliffe* v *Thackrah* the House of Lords reversed its own previous judgment and disposed of the myth that in the role of certifier the contract administrator is operating in a 'quasi judicial' role, and is immune from suit. The position is less clear with respect to the contract administrator's duty of care to the contractor. A certifier was found liable to a contractor in the case of *Michael Salliss & Co.* v *Calil and W. F. Newman & Associates*. Although this appeared to be overtaken in *Pacific Associates Inc* v *Baxter* the latter involved non-standard clauses which, if they had not been present, may have resulted in a different outcome.

Sutcliffe v Thackrah (1974) 4 BLR 16 (CA)

A contract administrator issued certificates on a contract for the construction of a dwelling house. The contractor's employment was determined for proper reasons following which the contractor went bankrupt. It then became apparent that much of the work, which had been included in the Interim Certificates, was defective, and the contract administrator was found negligent. In the House of Lords, when reviewing the role of the contract administrator, Lord Reid stated:

> Many matters may arise in the course of the execution of a building contract where a decision has to be made which will affect the amount of money which the contractor gets ... the building owner and the contractor make their contract on the understanding that in all such matters the contract administrator will act in a fair and unbiased manner and it must therefore be implied into the owner's contract with the contract administrator that he shall not only exercise due skill and care but also reach such decisions fairly holding the balance between his client and the contractor. (at page 21)

Michael Salliss & Co. Ltd v Calil and William F. Newman & Associates (1987) 13 ConLR 69

Calil employed contractor Michel Salliss for some refurbishment works on JCT63. W. F. Newman acted as contract administrator and quantity surveyor under the contract. The contractor commenced proceedings against the employer and joined the contract administrator as second defendants, claiming that the contract administrator was in breach of its duty to use all professional skill and care in granting only a 12-week extension of time when a 29-week extension was due. There was a sub-trial as to whether the contractor could recover damages against the contract administrator. His Honour Judge Fox-Andrews held that under a JCT contract the contract administrator owed a duty to the contractor to act fairly between the employer and contractor in matters such as certification and extensions of time. He also noted that:

> in many respects a contract administrator in circumstances such as these owes no duty to the contractors. He owes no duty to contractors in respect of the preparation of plans and specifications or in deciding matters such as whether or not he should cause a survey to be carried out. He owes no duty of care to a contractor whether or not he should order a variation. Once, however, he has ordered a variation he has to act fairly in pricing it. (at page 79)

Pacific Associates Inc. v Baxter (1988) 44 BLR 33 (CA)

Pacific Associates contracted on the FIDIC form of contract to carry out dredging and reclamation works for the Ruler of Dubai. The defendant was employed as engineer to administer the contract. Disputes arose between the employer and contractor which went to arbitration, and were subsequently settled with a £10 million payment to the contractor, with both parties paying their own costs. The contractor then brought a claim against the engineers for negligent certification, claiming the unrecovered balance of the claim together with interest and arbitration costs. His Honour Judge John Davies dismissed the claim. The contract contained a particular condition which stated:

> Neither any member of the Employer's staff nor the Engineer nor any of his staff, nor the Engineer's Representative shall be in any way personally liable for the acts or obligations under the Contract, or answerable for any default or omission on the part of the Employer in the observance or performance of any of the acts, matters or things which are herein contained.

The judge stated that 'the clear intention of (this clause) ... was to relieve the engineer of all personal liability for his acts and obligations under the contract' (13 ConLR 80, at page 93). He also stated that he felt the question of liability always depended on the particular terms of the contract in question, and that 'the over-riding intention of the contract was to put the engineer beyond the reach of legal responsibility for his acts' (at page 92). The contractor appealed but the appeal was dismissed. The Court of Appeal stated that the existence of the special clause meant that a duty of care could not in this case be imposed, but emphasised that otherwise such a duty might have existed.

8.3   The payment provisions in SBC05 have been drafted to ensure compliance with the Housing Grants, Construction and Regeneration Act 1996 (HGCRA 1996). Payment is to be made by the employer to the contractor after the work has been valued and following issue of certificates by the contract administrator (cl 4·9·1). This will normally be at monthly intervals, unless a different period is entered in the Contract Particulars.

8.4   SBC05 includes an optional provision for advance payment of the main contractor, and an entry must be made in the Contract Particulars to show whether or not it is to apply (cl 4·8). If it is, the amount will be entered as either a fixed sum, or a percentage of the contract sum. The entry must also show when it is to be paid to the contractor, and when it is to be reimbursed to the employer. A bond may be required, and SBC05 includes a form of advance payment bond in Schedule 6. There is always a risk in making an advance payment with respect to a construction contract, even when backed by a bond, and the procedure will inevitably involve extra expense to the employer. The employer should be quite clear as to what compensatory benefits, such as a reduction in the contract sum, will result before

Figure 21 Procedure at Interim Certificate

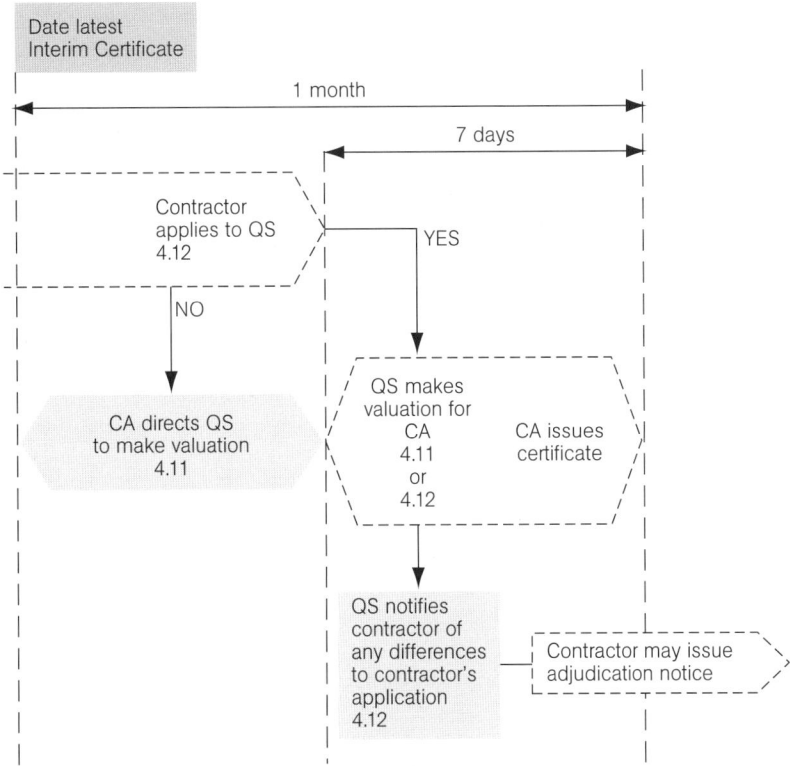

agreeing to any arrangement of this sort. In the end it is the employer's decision, but the contract administrator may need to explain the provisions and give initial advice.

Valuations and ascertainment of amounts due

8.5   The amount due to the contractor at interim payments is to be calculated as set out in clause 4·10. There are two methods by which it can be assessed: either through acceptance of the figure stated in a contractor's application for payment, or through valuation by the quantity surveyor (see Figure 21). The latter has traditionally been the basis for assessing payments under JCT forms.

8.6   Clause 4·12 gives the contractor the right to submit its own assessment of the value of an Interim Certificate. The application must be made at least seven days before the certificate is due and should be sent direct to the quantity surveyor. It is then open to the quantity surveyor to agree or disagree with this valuation; if the quantity surveyor disagrees with the amount he or she must identify in detail the matters where the differences arise. The contract administrator should request that the quantity surveyor forwards copies of all correspondence and keeps the contract administrator informed regarding applications.

8.7   If no application is received, or if the application is rejected, the valuation is carried out by the quantity surveyor as directed by the contract administrator and should cover the amounts due at a date not more than seven days before the issue of the certificate (cl 4·16). Valuations can be made 'whenever the contract administrator considers them necessary' except when the fluctuation formula adjustment procedure is used, where timing cannot be at the discretion of the contract administrator (cl 4·11). The roles of the quantity surveyor and the contract administrator, however, are quite distinct, and the valuation figure will not necessarily be the same as that on the Interim Certificate.

Coverage of the certificate

8.8   Clause 4·9·1 requires that Interim Certificates state not only the amount to be paid, but also 'to what the amount relates and the basis on which that amount was calculated'. It is unlikely that a great deal of detail will be required here; a short schedule will probably be sufficient. Similar provisions are included for the Final Certificate.

8.9   What the contract administrator is to include in Interim Certificates can be summarised as follows:

- total value of work properly executed (cl 4·16·1·1);

- total value of materials and goods properly on site (cl 4·16·1·2);

- value of off-site materials, goods or items, provided they are 'listed items' (cl 4·16·1·3).

8.10   There may also be amounts for work in connection with insurance claims, loss and expense and additional sums payable by the employer to the contractor for fees and charges, inspection tests, royalties, insurance, etc., as provided for by the contract (cl 4·16·2). There may also be deductions as set out in clause 4·16·3, for example where the employer has agreed to accept defective work. The total reached through the calculation set out in clauses 4·16·1 to 4·16·3 is referred to as the 'gross valuation'.

Value of work properly executed

8.11   The contract administrator should only certify after having carried out an inspection to a reasonably diligent standard. Contract administrators should not include any work that appears not to have been properly executed, whether or not it is about to be remedied or the retention is adequate to cover remedial work (*Townsend* v *Stone, Sutcliffe* v *Chippendale & Edmondson*). If in doubt, the contract administrator may require 'reasonable proof' from the contractor that materials and goods comply (cl 2·3·4) and may carry out, or request tests (cl 3·17). However SBC05 makes it clear that Interim Certificates are not conclusive evidence that work is in accordance with the contract. Where work, which has been included in a certificate, subsequently proves to be defective the value can be omitted from the next certificate.

8.12   The value of the work will be calculated using the rates shown in the bills or the Priced Document, whichever is appropriate. If a priced Activity Schedule is included, the amount shown on any Interim Certificate in respect of any items listed in the Activity Schedule should represent 95 per cent of the total of the amounts reached by multiplying the percentage of the work properly executed by the price for that work as shown on the Activity Schedule (cl 4·16·1·1). The fact that an Activity Schedule is used, however, does not lessen the contract administrator's duty to determine that all work certified has been carried out in accordance with the contract.

Townsend v Stone Toms & Partners (1984) 27 BLR 26 (CA)

Mr Townsend engaged contract administrators Stone Toms in connection with the renovation of a farmhouse in Somerset. John Laing Construction Ltd were employed to carry out the work on JCT67 Fixed Fee Form of Prime Cost Contract. Following the end of the rectification period the contract administrators issued an Interim Certificate that included the value of work which they had already included in their schedule of defects, and which they knew had not yet been put right. Mr Townsend brought proceedings against both Stone Toms and Laing. Laing made a payment into court of £30,000, which was accepted by Townsend in full and final settlement.

Townsend then continued with the proceedings against the contract administrator, claiming that they were entitled to recover any excess that they might have obtained for Laing had they continued with those proceedings. The official referee assessed the total value of the claims against Laing as only £25,000; therefore no excess was recoverable. The deputy official referee also found that the contract administrator was not negligent in issuing the Interim Certificate. Mr Townsend appealed and the Court of Appeal, although approving the lower court's decision on the effect of the payment into court, held that the contract administrator had been negligent. Oliver LJ stated:

> the whole purpose of the certification is to protect the client from paying to the builder more than the proper value of the work done, less proper retention, before it is due. If the contract administrator deliberately over-certifies work which he knows has not been done properly, this seems to be a clear breach of his contractual duty, and whether certification is described as 'negligent' or 'deliberate' is immaterial. (at page 46)

Sutcliffe v Chippendale & Edmondson (1971) 18 BLR 149

(Note that this case is the first instance decision which was appealed to the Court of Appeal sub nom *Sutcliffe* v *Thackrah*, discussed in paragraph 8.2 above.)

Mr Sutcliffe engaged the contract administrators, Chippendale & Edmondson, in relation to a project to build a new house. No terms of engagement were agreed but the contract administrators proceeded to design the house, invite tenders, and arrange for the appointment of a contractor on JCT63. Work progressed slowly and towards the end of the work it became obvious that much of the work was defective. The contract administrators had issued ten Interim Certificates before Mr Sutcliffe entirely lost confidence, dismissed the contract administrators and threw the contractor off the site. He then had the work completed by another contractor and other consultants, which cost around £7,000, in addition to which he was obliged, as a result of the original contractor having obtained judgment against him, to pay all ten certificates in full. As this contractor then went bankrupt he then brought a claim against the contract administrators.

The contract administrators contended, among other things, that their duty of supervision did not extend to informing the quantity surveyor of defective work that should be excluded from the valuation. His Honour Judge Stabb QC found for Mr Sutcliffe, stating 'I do not expect that the words "work properly executed" can include work not then properly executed but which it is expected, however confidently, the contractor will remedy in due course' (at page 166).

Unfixed and off-site materials and goods

8.13   The Interim Certificate should include materials which have been delivered to the site but not yet incorporated in the Works (cl 4·16·1·2), even though a limited risk to the employer remains (see paragraph 5.42). Clause 4·16·1·2, however, states that the obligation does not extend to materials that are prematurely delivered, or not properly protected. Contract administrators should pay careful attention to the exact wording of this proviso.

8.14   Under the SBC05 provisions the contract administrator is obliged to include the unfixed materials in an interim certificate, even though this limited risk to the employer remains. Clause 4·16·1·2, however, states that the obligation does not extend to materials that are prematurely delivered, or not properly protected. Contract administrators should pay careful attention to the exact wording of this proviso.

'Listed items'

8.15   Interim certificates might include amounts in respect of any 'listed items' (cl 4·16·1·3). These may be items 'uniquely identified' or not uniquely identified (i.e. materials or goods or items prefabricated for inclusion in the Works) and listed by the employer. If these provisions are to apply then the list must be attached to the bills of quantities (or specification/schedules of work) to which the contractor has tendered (cl 4·17). The value of items listed may be included in an Interim Certificate prior to delivery on site (cl 4·17) provided certain pre-conditions are fulfilled:

- the listed items are in accordance with the contract (cl 4·17·1);

- the contractor has provided reasonable proof that the property is vested in it (cl 4·17·2·1);

- the contractor provides proof that the items are insured against Specified Perils until delivery on site (cl 4·17·2·2);

- the listed items are 'set apart' or clearly marked and identified (cl 4·17·3);

- if the item is 'uniquely identified', a bond has been provided if required in the Contract Particulars (cl 4·17·4 Schedule 6);

- if the item is not 'uniquely identified', a bond has been provided (cl 4·17·5 Schedule 6).

8.16   It appears that the contract administrator has no discretionary power to certify any off-site items, other than those that have been listed. This makes the position for both parties clear, in that only 'listed' off-site materials are to be certified. The contract administrator should therefore be careful not to include any unlisted off-site materials in any certificate, as only listed items are covered by clause 2·25.

Other items in the gross valuation

8.17   In addition to the value of work properly executed and of materials properly on the site, clause 4·16·2 lists out other amounts that must be included in the gross valuation for an Interim Certificate. For example, payments ascertained as due to the main contractor under clause 4·23 (direct loss and/or expense) are to be added to the contract sum (cl 4·16·2·2). Where adjustments to the contract sum have been made in accordance with the terms of the contract then these must be taken into account at the next Interim Certificate (cl 4·4). JCT Practice Note 26 'Valuation and certification for interim payments including variations' gives helpful advice on procedures to avoid financial strain on the contractor arising from under-valuation or delays in valuation.

Retention

8.18   Some of the items that must be included in the gross valuation are subject to retention (cl 4·16·1), of which half is released upon practical completion. The retention percentage is the amount inserted in the Contract Particulars or if no amount is inserted is three per cent (cl 4·20). The employer is trustee for the beneficiaries of the retention, i.e. the contractor (cl 4·18·1) and, except where the employer is a local authority, may be required under clause 4·18·3 to place it in a separate banking account. The contract administrator must issue a statement of the amount withheld with each Interim Certificate (cl 4·18·2). The clause allows the employer the benefit of any interest which accrues. The employer has the right to deduct from the retention sums due from the contractor, including sums due by right of set-off (cl 4·13·2).

8.19   Retention has frequently been a point of controversy in the past. A series of cases established that the employer is obliged to place the money in a separate account, even if the contract contains no express provision, provided that it requires the employer to hold the money as a trustee (see *Wates Construction* v *Franthom Property* and *Finnegan* v *Ford Sellar Morris* below). However, the court will not make an order to place money in a separate account following the insolvency of the employer (see *Mac-Jordan Construction* v *Brookmount Erostin* below). The contractor would have no special claim beyond that of an unsecured creditor. To be safe the contractor must insist, while the employer is solvent, that the money is placed in a separate account.

Wates Construction (London) Ltd v Franthom Property Ltd (1991) 53 BLR 23 (CA)

Wates entered into a contract with Franthom on JCT80 to construct a hotel in Kent. Clause 30·5·3 (requiring Franthom to place retention in a separate account) had been deleted, but otherwise the retention clauses were in all material respects the same as those in SBC05. Although requested to by Wates, Franthom refused to place the accrued retention of around £84,000 in a separate account. Wates then commenced legal proceedings. Judge Newey ordered Franthom to place the money in an account, and Franthom then appealed. The court dismissed the appeal stating that 'clear express provisions are needed if a separate bank account is not to be set up'. The fact that the clause had been deleted did not of itself indicate what the parties' intentions were; the effect was the same as if the words had never been there at all.

J. F. Finnegan Ltd v Ford Sellar Morris Developments Ltd (1991) 53 BLR 38

Finnegan were contractors on a JCT81 contract for works at Ashford. After the Works reached practical completion the employer claimed liquidated damages of around £60,000 against a sum admitted as due to the contractor of around £20,000. Under clause 30·4·2·2 the employer was obliged to place retention monies deducted in a separate account if requested by the contractor. Finnegan commenced action to recover the sum due. The employer counter-claimed for the liquidated damages and Finnegan then requested that the retention be placed in a separate account. The employer refused and Finnegan applied for an injunction. The judge granted the injunction, despite the fact that this was long after practical completion. The contract did not require that a request was made each time retention was deducted nor at the time it was deducted.

Mac-Jordan Construction Ltd v Brookmount Erostin Ltd (1991) 56 BLR 1 (CA)

A developer held over £100,000 for the contractor in retention money but was also heavily indebted to the bank (floating loan granted by a charge). The developer went into insolvency and the bank appointed administrative receivers. The contractor then sought a court injunction to establish a separate retention fund but the Court of Appeal refused on grounds that this would give an unsecured creditor (the contractor) preference over any other unsecured creditors of an insolvent debtor. The contractor's right to the retention was stated to be no more than an 'unsatisfied and unsecured contractual right for the payment of money' (Scott LJ at page 15).

Bond in lieu of retention

8.20   This dilemma over retention and the effectiveness of trustee status has raised the question of bonds and guarantee bonds from both employer and contractor, respectively, as an alternative, and SBC05 now makes provision for the use of a bond in lieu of retention (cl 4·19). If a bond is to be required, then the Contract Particulars must indicate that clause 4·19 is to apply, and specify the maximum aggregate sum to be secured. The form of bond to be used is included in Schedule 6, Part 3, and must be provided by the contractor prior to the date of possession.

8.21   Where clause 4·19 applies, retention is not deducted from amounts on certificates. Instead, a statement of the retention that would have been deducted is prepared prior to each Interim Certificate. If at any time this statement exceeds the maximum aggregate sum stated in the bond, either the contractor arranges for the bond to be adjusted, or the employer deducts retention for the unsecured amount. If the contractor fails to provide the bond at all, the employer may deduct retention as described above.

Advance payments and bonds

8.22   The advance payment indicated in the Contract Particulars is paid to the contractor before the first certificate of payment is due for issue, but only after the contractor has provided the bond required (cl 4·8). Payment is made direct from the employer to the contractor, and the contract administrator should ensure receipt of copies of any correspondence regarding this. Details of when the reimbursements are to take place will also be set out in the Contract Particulars and could, for example, be in stages throughout the project. The reimbursement is deducted from the gross valuation under the relevant certificate (cl 4·10). It is not clear why clause 4·10 requires reimbursement of advance payments to be shown on certificates, as the original payment would not appear.

VAT

8.23   The contract sum is exclusive of any VAT (cl 4·6·1), and there is no requirement to indicate the VAT applicable to any certified amount. VAT is not a matter of contract but of statute and is normally paid by the employer on submission of a VAT invoice by the contractor following each Interim Certificate.

Payment procedure

8.24   The final date for payment of each Interim Certificate is 14 days from the date of issue (cl 4·13·1). Clauses 4·13·3 to 4·13·4 set out requirements for giving notice with regard to Interim Certificates. These clauses are required by the Housing Grants, Construction and Regeneration Act 1996 (HGCRA 1996) (ss 110 and 111), and are repeated with respect to the Final Certificate. The employer must give the contractor notice of how much it intends to pay no later than five days after an Interim Certificate is issued (cl 4·13·3). The clause follows the wording of the Act in that the notice is required whether or not the employer intends to make any deduction (see Figure 22). The guidance notes issued with Amendment 18 to JCT80, however, suggested that the employer would not have to give written notice if no deduction was intended.

8.25   If the employer does intend to withhold any amount, then it must give written notice of this no later than five days before the final date for payment, clearly stating the grounds for making the deduction, and the amount of deduction attributable to each ground (cl 4·13·4). The contract then states 'Subject to any notice given under clause 4·13·4 the Employer shall no later than the final date for payment pay the Contractor the amount specified in the notice under clause 4·13·3, or in the absence of a notice under 4·13·3, the amount stated as due in the Interim Certificate' (cl 4·13·5). It is suggested that if it is known in advance that a deduction is intended then the first notice must make this clear. The second notice may then be superfluous, always provided that the first notice included all the necessary information required under clause 4·13·4. If not all information is given, or new circumstances develop following the first notice stage, then the second notice will be required. In any event many employers may prefer to adopt the cautious and protective line of issuing both notices whenever a deduction is intended.

Deductions

8.26   The contract expressly gives the employer the right to make certain deductions from the certified sums due to the contractor, and this is clearly stated under clause 4·13·2. Notice of the deduction should have been given as described above. Deductions authorised by

Figure 22 Payment procedure

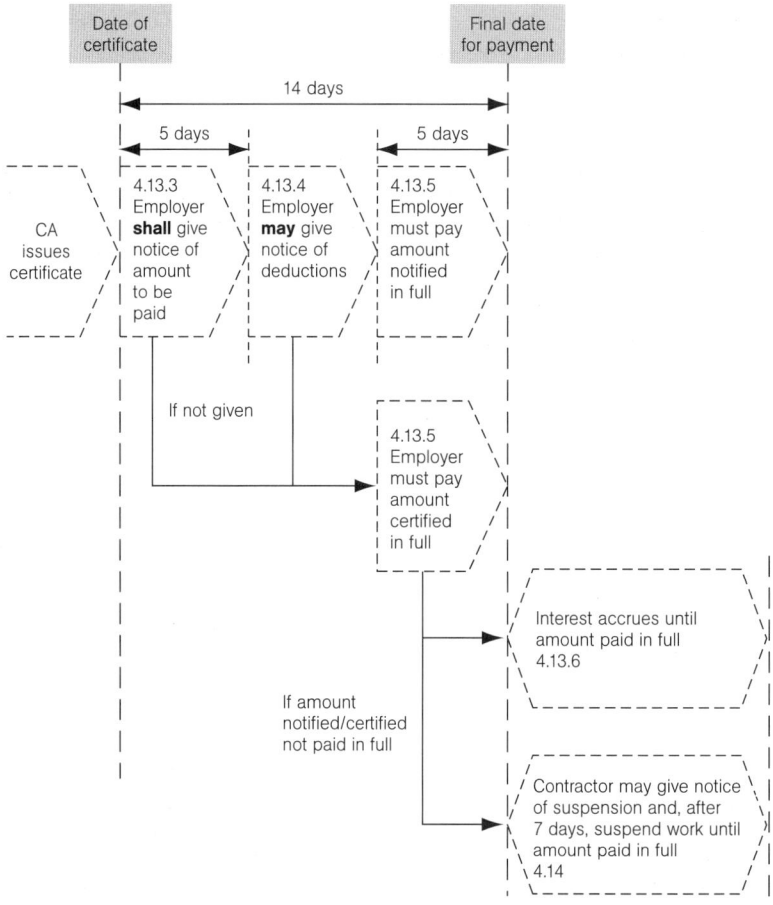

the contract arise in respect of:

- non-compliance with contract administrator's instructions (cl 3·11);

- default in taking out or maintaining insurance premium (cl 6·4·3 and *Schedule 3:A·2*);

- payment or allowance of liquidated damages (cl 2·32·2).

Employer's obligation to pay

8.27   In addition to the contractual rights to make deductions discussed above, the employer may have other rights to withhold payment under common law. Prior to the Housing Grants, Construction and Regeneration Act 1996 (HGCRA 1996) it was clear that if the employer had an arguable case that the certificate may have included work which was defective, and therefore had been over-valued, then the employer need not have paid the full amount, but could have raised the losses due to the defects either as a counterclaim in any action brought by the contractor, or as a defence to the claim (*Gilbert-Ash* v *Modern Engineering*, *Pillings* v *Kent*). The latter process is often termed 'abatement' by lawyers.

8.28   It is now generally agreed that in cases where such a right may exist, it can only be exercised through the use of the 'withholding notice' procedure as discussed above. The employer would therefore be unable to withhold amounts to cover any defective work included in a certificate, unless the deduction is covered by a notice (*Rupert Morgan Building Services* v *David Jervis*). The rights of the employer when defects appear after the expiry of the time limits for notices, but before the final date for payment, are unclear, but it is arguable that in such situations the employer would retain a right to abatement of the amount due.

Gilbert-Ash (Northern) Ltd v Modern Engineering (Bristol) Ltd (1974) 1 BLR 73 (HL)

Bradford City Council employed Gilbert-Ash as main contractor under JCT63. Modern Engineering was the sub-contractor for steelwork on a non-standard sub-contract drafted by Gilbert-Ash. The main contract stated that the main contractor should pay to the sub-contractor any amounts stated by the contract administrator as due to the sub-contractor under an Interim Certificate. Clause 14 of the sub-contract stated: 'if the Sub-contractor fails to comply with any of the conditions of this Sub-contract the Contractor reserves the right to suspend or withhold payment of any monies due or becoming due to the Sub-contractor'.

The contract administrator issued three Interim Certificates certifying that the total amount due in respect of the sub-contractor's work was £14,532 18s 10p, but out of that sum the contractor paid the sub-contractor only £10,000. Modern Engineering issued a writ claiming the balance. The main contractor issued a defence which asserted that it had claims against the sub-contractor for losses caused by delay and defective work. The question in issue was whether the main contractor had bound itself to pay sums in Interim Certificates without any right to set-off claims in respect of breaches of contract by the sub-contractor. The judge at first instance decided in favour of the contractor, and the sub-contractor appealed. The Court of Appeal reversed the decision, and the dispute was taken to the House of Lords. The House stated that the common law right of a defendant to set up a breach of warranty in diminution or extinction of an instalment of a sum which had become due could only be excluded by clear, unequivocal words. There was no provision in the main contract that excluded this right.

C. M. Pillings & Co. Ltd v Kent Investments Ltd (1985) 30 BLR 80 (CA)

The defendants employed the plaintiffs for work to a house on the JCT Fixed Fee Form of Prime Cost Contract. Following practical completion the contract administrator issued a certificate for over £100,000 which was not paid, and the contractor commenced legal proceedings, applying for summary judgment under Order 14. The defendant applied for a stay so that the matter could go to arbitration. The trial judge stayed the action and the contractor appealed. The appeal was dismissed. The court found that the effect of the terms of the contract was not that the certificate gave the contractor the right to immediate summary judgment for the amount. As the commentators stated: 'the issue of a certificate creates a debt and the burden is then upon the debtor to show that the amount apparently due is not owing'. If an arguable case can be shown, the matter can be decided at arbitration and payment of the certificate is not a condition precedent to this taking place.

Rupert Morgan Building Services (CCC) Ltd v David Jervis and Hamlet Jervis [2003] EWCA Cir 1583 CA

A couple engaged a builder to have works done on their cottage, by means of a contract on the standard form published by the Architecture and Surveying Institute. The 7th Interim Certificate was for a sum of around £44,000 plus VAT. The clients accepted that part of that amount was payable but disputed the balance amounting to some £27,000. The builders sought summary judgment for the balance. The clients did not give 'a notice of intention to withhold payment' before 'the prescribed period before the final date for payment'. The builders argued that it followed, by virtue of s.111(1) that the clients 'may not withhold payment'. The clients maintained that it was open to them by way of defence to prove that the items of work which go to make up the unpaid balance were not done at all, or were duplications of items already paid or were charged as extras when they were within the original contract, or represented 'snagging' for Works already done and paid for. The Court of Appeal determined that in the absence of an effective withholding notice the employer has no right to set off against a contract administrator's certificate.

Contractor's position if the certificate is not paid

8.29   SBC05 includes several provisions which protect the contractor if the employer fails to pay the contractor amounts due. Clause 4·13·6 makes provision for simple interest on late payments of certificates. This is set at three per cent over the Base Rate of the Bank of England, unless the Contract Particulars state some other rate, and the interest accrues from the final date for payment until the amount is paid. Similar provisions are included for the Final Certificate and for amounts due under sub-contracts. If the employer makes a valid deduction following a notice, it is suggested that interest would not be due on this amount. The clause does not refer to the amount stated on the certificate but to 'the amount due to the Contractor under the Conditions', which would take into account valid deductions.

8.30   The contractor is also given a 'right of suspension' under clause 4·14. This right is required by the Housing Grants, Construction and Regeneration Act 1996 (HGCRA 1996). If the employer fails to pay the contractor by the final date for payment, the contractor has a right to suspend performance of all its obligations under the contract, which would include not only the carrying out of the work but, for example, could also extend to any insurance obligations. This right is stated to be 'subject to any notice issued pursuant to clause 4·13·4', which suggests that the contractor may not suspend work if a valid notice to withhold payment has been given by the employer. The contractor must have given the employer written notice of its intention to suspend work and stated the grounds for the suspension, and the default must have continued for a further seven days. The contractor must resume work when the payment is made. Under these circumstances the suspension would not give the employer the right to terminate the contractor's employment. Any delay caused by the suspension is a Relevant Event (cl 2·29·5) and a 'matter' in relation to direct loss and/or expense (cl 4·24·3).

8.31   The contractor has the right to terminate the contract if the employer does not pay amounts properly due (cl 8·9·1). The contractor must give notice of this intention, which specifies the default as required by the contract.

Certificate not issued or undervalued

8.32   The issue of a certificate is a condition precedent of the right of the contractor to be paid (*Lubenham* v *South Pembrokeshire District Council*). This case states that the contractor is only entitled to the sum stated in the certificate, even if the certificate contains an error, for example because it includes a wrongful deduction. The contractor's remedy is to request that the error is corrected in the next certificate, or to bring proceedings to have the certificate adjusted. There are exceptions to this rule where, for example, the employer has interfered with the issue of the certificate, in which case the contractor may be entitled to summary judgment for the correct amount.

Lubenham Fidelities and Investments Co. Ltd v South Pembrokeshire District Council (1986) 33 BLR 39 (CA)

Lubenham Fidelities were a bondsman who elected to complete two building contracts, both based on JCT63. The contract administrators, Wigley Fox Partnership, issued several Interim Certificates which stated the total value of work carried out, but also made deductions for liquidated damages and defective work from the face of the certificate. Lubenham protested that the certificates had not been correctly calculated, withdrew their contractors from the site and issued notices to determine the contract. Shortly after the Council gave notice of determination of the contract. Lubenham brought a claim against the Council claiming that their notices were valid and effective, and against Wigley Fox on the basis that their negligence had caused them losses.

It was held that the Council were not obliged to pay more than the amount on the certificate, and that whatever the cause of the under-valuation the correct procedure was not to withdraw labour, but to request that the error was corrected in the next certificate, or to pursue the matter in arbitration. Lubenham's claim against Wigley Fox failed because it had been the suspension of the Works rather than the certificates that had caused the losses, and because the contract administrators had not acted with the intention of interfering with the performance of the contract.

Interim payment on practical completion

8.33   An Interim Certificate is to be issued at the end of the period during which the Practical Completion Certificate is issued (cl 4·9·2). The effect of clause 4·20·3 is to release to the contractor half of the retention that has been deducted. The employer retains the right to deduct half the percentage from the outstanding amounts due under this certificate.

8.34   There would normally be no further payment certificates between the interim payment at practical completion, and the Final Certificate. However in some circumstances, for example when a claim for loss and/or expense has not been resolved prior to practical completion, the architect should issue a certificate as soon as the claim has been ascertained (cl 4·9·2), but would not be required to issue one within one calendar month of a previous certificate.

This would apply only to claims or disputed matters; there should be no work instructed or carried out which requires certification following practical completion.

Final Certificate

8.35   To summarise, by Final Certificate stage the following certificates should have been issued:

- Interim Certificates at monthly intervals (cl 4·9·2);

- Practical Completion Certificate (cl 2·30);

- Interim Certificate following practical completion, including release of half of the retention (cl 4·20·3);

- certificates during the rectification period (cl 4·9·2);

- Certificate of Making Good (cl 2·39).

8.36   The Final Certificate must be issued within the specific time periods set out in the contract (cl 4·15·1). In practice the latest date for issue tends to be determined by the process of calculating the adjusted contract sum. The onus is on the contractor to send all necessary information to the contract administrator or the quantity surveyor not later than six months after practical completion of the Works (cl 4·5·1). No later than three months after receiving this information, the contract administrator, or the quantity surveyor if asked to do so, makes a final assessment of the amount of loss and expense due (cl 4·5·2·1), and the quantity surveyor prepares a statement of final adjustment of the contract sum (cl 4·5·2·2). This must be sent to the contractor 'forthwith'.

8.37   The Final Certificate is then issued within two months of this statement (or, within two months of the issue of the last Certificate of Making Good, or the expiry of the last rectification period, whichever is the latest) (cl 4·15·1). It is worth noting that it has been held that the Final Certificate can be issued at the same time as the statement, although it would be good practice to allow the contractor time to consider the document (*Penwith District Council v V. P. Developments*). It is worth noting that a document does not necessarily have to be headed 'Final Certificate, or be in any particular format, to comply with clause 4·1·5: whether it constitutes a Final Certificate will depend on the facts in each case (*B. R. Cantrell (2) E. P. Cantrell v Wright and Fuller Ltd*).

Penwith District Council v V. P. Developments Ltd 21 May 1999, unreported

Penwith employed V. P. for maintenance works to 91 houses at Hayle. The contract was on JCT80. Practical completion took place on 21 September 1990, and the Certificate of Making Good Defects was issued on 30 October 1991. V. P. submitted a draft final account on 14 January 1991. Three interim certificates were issued following practical completion, the last one on 10 July 1992. The Final Certificate was issued on 8 April 1993, and enclosed a document summarising how the figure on the Final Certificate had been arrived at. V. P. gave notice of arbitration some three years later. It argued that it was not barred by the clause 30·9 conclusiveness provisions as the Final Certificate had not been valid.

The arbitrator found for V. P., stating that the intention of the contract was that the contractor should have at least three months to consider the ascertainment of final account referred to in clause 30·6·1. Penwith appealed and His Honour Judge Humphrey Lloyd found that the contract terms required that no minimum period should have elapsed, all the time limits referred to were maxima. He also found that no such term could be implied: 'the 1980 JCT form is a long and complex document and was plainly intended to provide for most conceivable circumstances and to block the many attempts to find gaps in its structures, despite repeated assaults'.

8.38   The Final Certificate must state the contract sum as adjusted under clause 4·3, which sets out all the deductions and additions to the contract sum (cl 4·15·2). As with interim certificates, the Final Certificate can be for a negative amount – in other words it can certify that payment is due from the contractor to the employer. The final date for payment of the Final Certificate is 28 days from the date of issue, and subject to equivalent notices provisions as described above in relation to Interim Certificates (cl 4·15·3–4·15·7).

Conclusive effect of Final Certificate

8.39   The Final Certificate is conclusive evidence that proper adjustment has been made to the contract sum (cl 1·10·1·2), and the contractor is prevented from seeking to raise any further claims for extensions of time (cl 1·10·1·3), or for reimbursement of direct/loss and/or expense (cl 1·10·1·4). It is also conclusive evidence that where matters have been expressly stated to be for the approval of the contract administrator they have been approved (cl 1·10·1·1), but apart from those matters it is not conclusive that any other materials, workmanship, etc. are in accordance with the contract.

8.40   Both parties have the right to challenge the issue of the Final Certificate by commencing proceedings within 28 days. The certificate is then only conclusive with respect to matters that are not challenged in the proceedings (cl 1·10·3). Once a party has taken the initial steps, but then takes no further action for a period of 12 months, then the certificate has the conclusive effect stated (cl 1·10·2·2). The bar on raising matters after the 28-day period cannot be extended by the court, as the bar is an evidential bar and not a bar to bringing arbitration proceedings. In other words arbitration can be commenced, but no evidence can be brought forward. In cases where a dispute has been raised in adjudication, and the adjudicator's decision is reached after the Final Certificate, if either party wishes to challenge that decision then it must initiate proceedings within 28 days of the date of the decision (cl 1·10·2·4).

8.41   The conclusive effect of the Final Certificate was the subject of much heated debate around 10 years ago, following the decisions in *Colbart Ltd* v *H. Kumar* and *Crown Estate Commissioners* v *John Mowlem*, cases which would still apply to older versions of JCT forms (see *London Borough of Barking & Dagenham* v *Terrapin*). Following the cases the JCT amended the relevant clauses in its contracts. It appears unlikely, under the current wording, that an employer would be unable to effectively raise a claim regarding work or materials which were not in accordance with the contract following the 28-days cut-off period, provided that it had not been stated to be 'to approval' of the contract administrator somewhere within the contract documents.

8.42   In cases where work has been stated to be 'to approval' the contract administrator should take particular care, and should not issue the certificate unless and until the work is satisfactory. Where unsatisfactory work has been accepted the contract administrator should, together with the employer, weigh the advantages of issuing the Final Certificate (the element of finality it brings to matters such as the final account) against the disadvantages (not being able to claim with respect to the unsatisfactory work), before deciding whether to proceed.

Crown Estate Commissioners v John Mowlem & Co. Ltd (1994) 70 BLR 1 (CA)

Crown Estates employed Mowlem to construct a commercial development on the site of the former Kensington Palace Barracks. A Final Certificate was issued on 2 December 1992, and on 6 April 1993 Crown Estates gave notice of arbitration. They then issued a summons under s.27 of the Arbitration Act for an order extending the time within which to commence arbitration, in order to validate their notice. In addition to the summons the judge at first instance was also asked to consider the question as to what, if anything, the Final Certificate was conclusive evidence of, as this would affect what could be raised in the arbitration.

The judge issued the order extending time and held that the Final Certificate was only conclusive as to matters that were expressly stated to be for the satisfaction of the contract administrator. Mowlem appealed and the appeal was allowed. The Court of Appeal stated that clauses 30·9·1·1 and 30·9·3 did not limit the time within which arbitration proceedings could be brought, therefore the court had no powers under the Arbitration Act that could defeat the effect of the certificate. It also held that, as all standards and quality of work and materials were inherently matters for the opinion of the contract administrator, the Final Certificate was conclusive evidence of all such matters.

London Borough of Barking & Dagenham v Terrapin Construction Ltd [2000] BLR 479

The Borough employed Terrapin Construction to design and build new and refurbishment work at a school in Dagenham. No document entitled 'Employer's Requirements' had been issued to the contractor at tender stage, but the contractor had been given a 'brief' which set out in general terms the nature of works which the Borough wanted designed, and the court decided requirements were 'represented by the contract as a whole'. The contract was to be on WCD81. Once a tender figure had been negotiated the Borough sent the contractor an order for the work which set out the agreed contract figure, incorporated the terms of WCD81, and also stated 'In consideration of this Agreement hereinafter contained on the part of the employer the Contractor shall and will execute complete and maintain the Works in all respects to the satisfaction of the Controller of Development and Technical Services'. The court decided that in this context the Final Statement was conclusive evidence that all work had been carried out to the satisfaction of the employer.

B. R. Cantrell (2) E. P. Cantrell v Wright and Fuller Ltd [2003] BLR 412

Cantrell engaged Wright on a JCT80 form to construct an extension to a nursing home in Woodbridge, Suffolk. The work achieved practical completion on 23 February 1998. Following practical completion the contract administrator did not issue an extension of time, nor had he issued a Certificate of Non-completion. Following meetings between quantity surveyors, a document entitled 'final account' was agreed. On 12 March 1999 the contract administrator sent the claimant the final account plus an Interim Certificate to the sum of around £25,000. On 29 March 1999 the contract administrator issued a further certificate, which referred to a 'final payment', and was accompanied by a letter which referred to it as a 'final certificate'. The employer's solicitors immediately challenged the adjustments to the contract sum, and the contractor's solicitors demanded payment. In May 2002 a notice of arbitration was served and an arbitrator appointed. The parties were in dispute as to whether the March 1999 document was a Final Certificate complying with clause 30·8. If it was, then its conclusive effect would defeat some of the matters claimed. The arbitrator decided that the certificate was a Final Certificate.

9 Insurance

9.1 Construction operations can be hazardous, and it is therefore important that liability for losses resulting from personal injury, damage to any property, or to the Works, is clearly allocated to one party or the other, and that the liability is backed up by insurance.

9.2 It is usual for the contractor to indemnify the employer in respect of certain losses, for example for injury to persons, or damage to neighbouring property which has been caused by the contractor's negligence, and in SBC05 this is done under clauses 6·1 and 6·2. This indemnity protects the employer in that if an injured party brings an action against the employer, rather than against the contractor, the latter has agreed to carry the consequences of the claim. If a third party sues the employer, then the employer can join the contractor as co-defendant or bring separate proceedings. Indemnities given to the employer by the contractor will obviously be quite worthless unless there are adequate resources to meet claims. The contract therefore requires insurance cover under clause 6·4 to back up the indemnities.

9.3 In addition to the requirement for insurance against claims arising in respect of persons and property, the contract also contains alternative provisions for insurance of the Works. There is an optional provision for insurance against damage caused to property which is not the result of the negligence of the contractor, if required by the employer.

Injury to persons and damage to property

9.4 Clauses 6·1–6·3 cover injury to persons and damage to property other than the Works, which arise from the carrying out of the Works. The contractor is required to match the indemnities given in clauses 6·1 and 6·2 with insurance under clause 6·4. The contractor must be able to provide evidence that this insurance has been taken out (cl 6·4·2). The minimum cover required as a contractual obligation is entered in the Contract Particulars. If the contractor defaults, the employer may take out the insurance and deduct the cost from the contract sum (cl 6·4·3).

9.5 The contractor's liability in respect of personal injury or death of employees is met by an employer's liability policy. This has been compulsory since the Employer's Liability (Compulsory Insurance) Act 1969. Clause 6·4·1·2 covers the insurance required by this Act. The clause refers to 'all relevant legislation', which would also cover, for example, insurance requirements under the Road Traffic Act.

9.6 The contractor's liability in respect of third parties (death or personal injury and loss or damage to property including consequential loss) is met by its public liability policy. The contractor is required to insure the indemnities required under clause 6 up to the amount stated in the Contract Particulars (cl 6·4·1·2). Liability at common law for claims by third parties is unlimited, and any amount specified in the contract is merely the employer's requirement in the interests of safeguarding against inadequacies, and in no way limits the contractor's liability under clause 6·1. It is recognised in footnote [50] to clause 6·1·2 that it may not always be possible to acquire insurance cover which is co-extensive with the indemnity required in clauses 6·1 and 6·2. For example, the insurance market has removed gradual pollution from its public liability policies. This again does not affect the contractor's duty to indemnify.

9.7   The liability and duty to indemnify are subject to exceptions. In respect of liability for personal injury or death, this is qualified in that the contractor is not liable where injury or death is caused by an act of the employer, or a person for whom the employer is responsible (cl 6·1).

9.8   In respect of damage to property, the contractor is only liable to the extent that the damage is caused by negligence or breach of statutory duty or other default of the contractor or the contractor's persons (cl 6·2). The contractor is therefore liable only for losses caused by its own defaults. It is made clear in clause 6·3·1 that the definition of 'property' excludes the Works, up to practical completion. However the Contractor is liable for any damage it negligently caused to parts taken over by partial possession (cl 6·3·3), or to a section of the works, after the relevant section Completion Certificate is issued (cl 6·3·2) or to the Works after practical completion. The last sentence of clause 6·2 excludes liability for loss or damage to property caused by a 'Specified Peril' where this is required to be insured under Option C. This means that, where Option C is applicable, the contractor is not liable for losses insured under that clause and caused by a 'Specified Peril', even where the damage is caused by the contractor's own negligence (*Scottish Special Housing Association* v *Wimpey Construction, Kruger Tissue (Industrial) Ltd* v *Frank Galliers Ltd* (1998) 57 ConLR 1, *Cooperative Retail Services Limited* v *Taylor Young Partnerships* [2002] BLR 272). Domestic sub-contractors, however, may be liable for losses caused by their negligence (*BT* v *James Thompson & Sons*). It should be noted, also, that the contractor might remain liable for some consequential losses (*Kruger Tissue* v *Frank Galliers*).

Scottish Special Housing Association v Wimpey Construction UK Ltd (1986) 34 BLR 1

SSHA entered into a contract with Wimpey on JCT63 (1977 edition) to modernise 128 houses in Edinburgh. During the course of the Works some of the houses were damaged by fire and it was assumed for the purposes of the case that the fire was caused by the contractor's negligence. Clause 18(2) [the equivalent is 20·2 in SBC05], which dealt with the contractor's liability for damage to property, was headed by the phrase 'except for such loss and damage as is at the risk of the Employer under … 20(C) [22C in SBC05]'. The court found that this had the effect of exempting the contractor for liability for any loss or damage that was required to be insured under 20(C), however caused.

British Telecommunications plc v James Thompson & Sons (Engineers) Ltd [1999] BLR 35 (HL)

James Thompson were sub-contractors on a refurbishment project for British Telecom (BT). A fire broke out in the roof area while the sub-contractors were carrying out their work. The court found that the relevant clauses had the same effect as the equivalent clauses considered in *SSHA* v *Wimpey*.

However, it decided that domestic sub-contractors remained under a duty of care to prevent such losses, and were therefore liable to BT under the tort of negligence. It considered that the wording of clause 22·3, which required the joint names policies to waive the rights of subrogation against nominated but not domestic sub-contractors, should be taken into account in considering whether a duty of care existed. The fact that BT were indemnified by the clause 22C insurers, even if the fire was caused by the sub-contractors, was not sufficient to prevent the imposition of the duty.

Kruger Tissue (Industrial) Ltd v Frank Galliers Ltd (1998) 57 ConLR 1

Damage was caused to existing building and works by fire, assumed for the purposes of the case to be the result of the negligence of the contractor or sub-contractor. The employers brought a claim for loss of profits, increased cost of working and consultants fees, all of which are consequential losses. Judge John Hicks decided that the employer's duty to insure for 'the full cost of reinstatement, repair or replacement of the existing structure and the Works under clause 22C (and therefore contractors' exemption from liability under clause 20·2), did not include such consequential losses'. A claim could therefore be brought against the contractor for these.

Cooperative Retail Services Limited v Taylor Young Partnerships [2002] BLR 272

Cooperative Retail Services (CRS) engaged Wimpey construction to construct a new office block in Rochdale. Taylor Young Partnership (TYP) were the architects, and Hoare Lee and Partners (HLP) the mechanical and engineering consultants. In 1995, before practical completion, a fire occurred, which caused extensive damage to the incomplete building. CRS brought a claim against TYP and HLP, who joined in Wimpey and Hall (electrical sub-contractors). It was assumed for the purposes of the trial that the damage was caused by breach of contract of Wimpey, Hall, TVP and HLP. TVP and HLP would therefore be liable to CRS for losses due to damage caused, but would only be able to claim a contribution from Wimpey and Hall if it could be shown that they were also liable to CRS. The House of Lords (dismissing an appeal by TVP) determined that the combined effect of clauses 20·3 and 22A·1 was that Wimpey and Hall were not liable for this damage.

Scottish and Newcastle plc v G. D. Construction (St Albans) Ltd [2003] BLR 131

Scottish and Newcastle entered into a contract with GD on IFC84 to carry out refurbishment work to a public house in Reading. During the course of work a fire broke out. There was a trial of a preliminary issue as to whether the contractor was liable to the employer for the cost of the repair to the existing structure, and it was assumed for the purposes of the trial that the fire was caused by the contractor's negligence. The trial judge found the contractor liable and the contractor appealed. It was held, allowing the appeal, that the combined effect of clauses 6·1·2 and 6·3·1C was that the contractor was not liable for these losses. The express linkage between these two clauses was considered key to reaching this decision.

Damage to property not caused by the negligence of the contractor

9.9   The liability for damage to adjoining buildings where there has been no negligence on the part of the contractor is not covered under clause 6·2. Subsidence or vibration resulting from the carrying out of the Works might cause such damage, even though the contractor has taken reasonable care. This is a risk which may be quite high with certain projects on tight urban sites, or in close proximity to old buildings. In such cases it might be advisable to take out a special policy for the benefit of the employer.

9.10   In SBC05 there is an optional provision for this type of insurance under clause 6·5·1. If it is anticipated that the main contractor may be required to take out this insurance, the correct deletion must be made in the Contract Particulars, and the amount of cover entered. The contract administrator must then instruct the contractor to take out the policy, after confirming with the employer that the policy is required. The cost is added to the contract sum. The policy must be in joint names and placed with insurers approved by the employer. The policy and receipt are to be deposited with the employer (cl 6·5·2).

9.11   This insurance is usually expensive, and subject to a great many exceptions. If it is required, then the policy needs to be effective at the start of the site operations when demolition, excavation, etc. are carried out. The text of clause 6·5·1 was revised in 1996 to take account of the wording of model exclusions compiled by the Association of British Insurers. The policy should be checked by the employer's insurance advisers to ensure that any exclusions correlate with clause 6·5·1 and that the policy provides the cover that the clause requires.

Insurance of the Works

9.12   There are three alternative options for insuring the Works, which are set out in schedule 3 and the option which is applicable should be entered in the Contract Particulars. In all cases the policies are to be in joint names, and cover must be maintained up until practical completion of the Works, or termination if this should occur earlier. All three are for joint names policies. The 'Joint Names Policy' definition was reworded in 1996 to make clear the intention that, under the policy, the insurer does not have a right of subrogation to recover any of the monies from either of the named parties. The policies must also either cover sub-contractors or include a waiver of any rights of subrogation against them (cl 6·9·1). With the exception of joint names policies under Option C, the requirements for recognition or waiver apply also to domestic sub-contractors.

9.13   Options A and B are for insuring new building work and require 'All Risks' cover under joint names policies. A definition of 'All Risks' is given in clause 6·8 and refers to 'any physical loss or damage to work executed and Site Materials and against the reasonable cost of the removal and disposal of debris ... '. There is also a list of exclusions, which includes the cost necessary to repair, replace or rectify property which is defective, loss or damage due to defective design, and loss or damage arising from war and hostilities. Even in a so-called 'All Risks' insurance policy there may be further exclusions, and the employer's insurance advisers should carefully check the wording of each policy.

9.14   Option A insurance is taken out by the contractor and is to be for the full reinstatement value of the Works, including professional fees to the extent entered in the Contract Particulars (cl A·1). If no percentage is stated for professional fees, the default rate is 15%)

The contractor is responsible for keeping the Works fully covered, and must provide evidence as required by clause A·2 that the insurance has been taken out. If the contractor fails to take out insurance the employer may do so, and deduct the cost from monies due to the contractor (cl A·2).

9.15   Option B insurance is taken out by the employer, and again is to be for the full reinstatement value of the Works, including professional fees (cl B·1). The employer is responsible for keeping the Works fully covered and, except where the employer is a local Authority, must provide evidence that the policy has been taken out (cl B2·1). The contractor may take out the policy if the employer details and the amount is added to the Contract Sum.

9.16   Option C is applicable where work is being carried out to existing buildings. It includes two insurances, both taken out by the employer. The existing structure and contents must be insured against 'Specified Perils' as defined in clause 6·8 (cl C·1). New works in, or extensions to, existing buildings must be covered by an 'All Risks' insurance policy (cl C·2). Evidence of insurance must be provided, and the contractor has rights in default as described above (C 3:1). If the insurers of the works require measures to remedy any breach of the code, the party receiving notification must copy it to the other (cl 6·15·1).

Action following damage to the Works

9.17   The procedure is similar under Options A, B and C. The contractor must notify the contract administrator and the employer of the details of the damage (A4·1, B3·1, C4·1). The insurers are immediately informed. After any inspection required has been made by the insurers the contractor is then obliged to make good the damage and continue with the Works (A4·3, B3·3, C4·5). Under Option C either party is given the right to terminate the employment of the contractor 'if it is just and equitable to do so' (cl C·4·4·1), in which case the provisions of clause 8·12 will apply. All monies due under the insurance policy are paid direct to the employer (A4·4, B3·4, C4·3).

9.18   All three Options state that 'the occurrence of such loss or damage shall be disregarded in computing any amounts payable to the contractor' (A4·2, B3·2, C4·2). Interim certificates that have already been issued and the amounts paid or due under them are of course not affected by the occurrence of the damage. In addition, any work that was completed after the most recent Interim Certificate, but was then subsequently damaged, should also be included in the next Interim Certificate.

9.19   Under clause Option A the contractor must take out insurance for the full reinstatement value of the Works, plus a percentage to cover professional fees if this is required in the Contract Particulars (cl A·1). In the event of any damage occurring, the insurance money paid to the employer, minus the part of it to cover professional fees is paid through certificates issued by the contract administrator, 'issued on the dates fixed for the issue of Interim Certificates' (A4·4). This implies a separate series of certificates, distinct from the normal Interim Certificates, which are commonly headed in practice 're-instatement certificates'. If the amount paid by the insurers is less than it costs the contractor to rebuild the Works, the contractor is not entitled to any additional payment (cl A4·5). The risk of any under insurance therefore lies with the contractor. Under Options B and C the rebuilding work is treated as if it were a variation under clause 3·14, therefore the contractor is less at risk and the employer will have to bear any shortfall in the monies paid out.

9.20   Under clause 2·29·9 the contractor is entitled to an extension of time for delay caused by loss or damage due to one or more of the 'Specified Perils'. In addition, if the work is treated as a 3·14 variation, the contractor may be entitled to an extension of time and loss and/or expense under clauses 2·29·1 and 4·24·1. In all cases the entitlement appears to extend even to cases where the damage was caused by the contractor's negligence.

Terrorism Cover

9.21   'Terrorism Cover', is defined as insurance provided under Options A, B or C for physical loss or damage to the work or Site Materials, or to an existing structure and/or its contents, caused by terrorism. Although there have been difficulties in the past in obtaining such cover, at the time of publication of SBC05 the insurance market was prepared to cover such risks. If, however, the insurer named in the Joint Names policy decide to withdraw this cover and notify either party, it must immediately notify the other that Terrorism Cover has ceased (cl 6·10·1).

9.22   The employer must then decide whether or not it wishes to continue with the Works, and notify the contractor accordingly (cl 6·10·2). If the employer decides to terminate the contractor's employment, the provisions of 8·12·2 to 8·12·5 apply, excluding clause 8·12·3·5 (cl 6·10·3, see Chapter 10). Otherwise, should any damage be caused by terrorism, the contractor is required to make good the damage and the related work is treated as a variation (cl 6·10·4).

Professional Indemnity Insurance

9.23   The contractor is required to carry professional indemnity insurance (cl 6·11). The level and amount of cover must be inserted in the Contract Particulars – if no level is inserted it will be 'the aggregate amount for any one period of insurance', and if no amount is stated no insurance will be required. There is provision for inserting a level of cover for pollution or contamination claims. In addition if the expiry period is to be 12 years from practical completion then this must be indicated, or otherwise the period will be six years. The insurance must be taken out immediately following the execution of the contract, and maintained until the date of practical completion. The contractor must provide evidence of the insurance if required.

The Joint Fire Code

9.24   The Joint Fire Code (cl 6·14) is designed to reduce the incidence of fire on construction sites. It is an optional provision (cl 6·13), but as compliance with the Code may reduce the cost of some insurance policies, its inclusion should be carefully considered. If included, both parties undertake to comply with the Code and ensure that those employed by them also comply (cl 6·14). The contractor is required to carry out any remedial measures specified in a notice that relate to its obligation to complete the works (cl 6·15·1·1). If the contractor does not comply with the notice within seven days, the employer may employ and pay others to effect such compliance (cl 6·15·2). Where the specified measures require a variation to the works, the contract administrator should issue an instruction (cl 6·15·1·2).

9.25   There are some potential problems with respect to the Fire Code clauses. There is no absolute requirement that the remedial measures required by the insurers must be covered by an instruction of the contract administrator, something that could give rise to confusion in practice. Also, there is no reciprocal right for the contractor to employ others etc., if the employer fails in some way to comply with a requirement of the insurers. This might arise, for example, where persons engaged by the employer fail to comply.

Employer's loss of liquidated damages

9.26   Insurance against loss of liquidated damages is no longer covered by the contract, so any employer wishing to take out this insurance would have to make its own arrangements. There are problems with such insurance as liquidated damages are payable without proof, and traditionally insurers only pay on proof of actual loss. As a result, only one or two firms are currently willing to offer cover, and the price tends to be high.

Other insurance

9.27   There are other insurances not covered by the provisions of SBC05, which the employer might wish to consider. The employer is the party in the best position to assess possible loss. Where there are likely to be business or other economic losses, then these can be insured against, albeit at a price. It is also possible to insure against defects occurring in the buildings by means of project-related insurance. This is still relatively expensive and limited to a ten-year 'decennial' loss. Irrespective of blame, it means that money is available for remedying the defects which will occur most often in the first eight years of the life of a building. Project-related insurance needs to include for subrogation waiver, and in no way reduces the need for professional indemnity cover.

The contract administrator's role in insurance

9.28   The contract administrator has a duty to explain the provisions of the contract to the employer. The choice of the appropriate option for insuring 'the Works' is particularly important, and advice must be given to the employer concerning the consequences.

9.29   The employer should take advice from its own insurance experts concerning the suitability and wording of any policies. The contract administrator is primarily a channel of communication, and although a check should be carried out on wording to see that no undesirable exceptions or restrictions exist that might affect the carrying out of the Works, the main responsibility should rest with the employer and the employer's broker or insurance advisers.

9.30   Where the insurance requirements of the contract cannot be matched by effective cover, then the employer should seek expert advice. For example, the building might be special and uninsurable, or the employer might not wish to have insurance, etc. Decisions in such situations will also have implications for contractors and sub-contractors, and expert advice must be sought.

10 Termination

10.1 Despite the good intentions of both parties at the outset of a contract, breaches of contract sometimes occur. Some breaches are mere technicalities (for example, failure to issue a clause 4·13·3 notice when the employer intends to pay the certified amount in full), some are more serious, although it may be difficult to substantiate damage (for example, non-provision of 'master programme' to time); some are so serious that they go to the root of the contract.

10.2 Building contracts usually have express provisions to deal with certain foreseeable situations which might otherwise be breaches. For example, failure to give possession of the site can be dealt with under deferment (cl 2·5), failure to provide information at the agreed dates can be dealt with under extensions of time (cl 2·28), and imposing new restrictions on the contractor's working methods can be dealt with under variations (cl 5·1). These can all therefore be dealt with by the 'machinery' of the contract. If the machinery is not operated as it should be, then the injured party may be able to claim damages for breach of contract. Such claims would have to go to adjudication, arbitration or litigation. For more serious breaches, the contract contains provisions allowing the termination of the employment of the contractor.

Repudiation or termination

10.3 Where the behaviour of one party makes it difficult or impossible for the other to carry out the obligations of the contract, the injured party might allege prevention of performance and sue either for damages or a quantum meruit.

10.4 Where it is impossible to expect further performance then the allegation might be that of repudiation. Repudiation is when one party makes it clear that they no longer intend to be bound by the provisions of the contract. This might be expressly stated, or implied by the party's behaviour.

10.5 JCT contracts have traditionally included determination clauses (now 'termination'), which provide for the effective termination of the employment of the contractor in circumstances which may amount to, or which may fall short of, repudiation. It should be noted that the termination is of the contractor's employment, and is not termination of the contract. This means that the parties remain bound by its provisions. If there is repudiation, invoking a termination clause is unnecessary, because the injured party can accept the repudiation and bring the contract to an end. However, the termination provisions are useful in setting out the exact circumstances, procedures and consequences of the termination of employment. If the termination provisions are operated unjustifiably by either party this in itself could amount to a repudiation of the contract. This, in turn, might give the other party the right to treat the contract as at an end and claim damages.

10.6   Termination can be initiated by the employer (cl 8·4) in the event of specified defaults by the contractor such as suspending the Works or failing to comply with the CDM Regulations, or in the event of the insolvency of the contractor. Termination can be initiated by the contractor (cl 8·9) in the event of specified defaults by the employer such as failure to pay the amount due on a certificate, or where specified events result in the suspension of work beyond a period to be entered in the Contract Particulars. Termination might also follow the insolvency of the employer. In the event of neutral causes, which bring about the suspension of the uncompleted Works for the period listed in the Contract Particulars, termination can be exercised by either party (cl 8·11).

Termination by the employer

10.7   The contract provides for termination of the employment of the contractor under stated circumstances (cl 8·4). SBC05 expressly states that the right to terminate the contractor's employment is 'without prejudice to any other rights or remedies' (cl 8·3·1). This termination can be initiated by the employer in the event of specified defaults by the contractor occurring prior to practical completion (cl 8·4), the insolvency of the contractor (cl 8·5), or corruption (cl 8·6) (note that in the case of bankruptcy of the contractor, the termination is no longer automatic as it was under JCT98 clause 27·3·3).

10.8   The procedures as set out in the contract must be followed exactly, especially those concerning the issue of notices under clause 8·4 (see Figure 23). If default occurs the contract administrator should issue a warning notice (cl 8·4·1). If the default continues for 14 days from receipt of the notice, then the employer may terminate the employment of the contractor by the issue of a further notice within 10 days from the expiry of the 14 days (cl 8·4·2). If the contractor ends the default or if the employer gives no further notice, and the contractor then repeats the default, the employer may terminate 'within a reasonable time after such repetition' (cl 8·4·3). The employer must still give a notice of termination, but no further warning is required from the contract administrator. There appears to be no time limit on the repetition of the default. It should be noted that, to be valid, all notices must be in writing and given by actual delivery, or by special or recorded delivery (cl 8·2·3). This rules out e-mail or fax transmissions. As time limits are of vital importance it is also wise to have receipt of delivery confirmed.

10.9   The grounds for termination by the employer must be clearly established and expressed. The contract clearly states that termination must not be exercised unreasonably or vexatiously (cl 8·2·1). Before issuing any notice the contract administrator should check, for example, that all extensions of time have been dealt with in accordance with the contract. Under clause 8·4·1 suspension of the work must be whole and substantial, and 'without reasonable cause'. However, the contractor might find 'reasonable cause' in any of the matters referred to in clause 8·9·1. An exercise of the contractor's right to suspend work (HGCRA 1996) would not be cause for termination, provided that it had been exercised in accordance with the terms of the contract.

10.10   One of the defaults listed in clause 8·4·1 is that the contractor 'fails to proceed regularly and diligently'. This is notoriously difficult to establish, and although meticulous records will help, contract administrators are often understandably reluctant to issue the first

Figure 23 Termination by employer

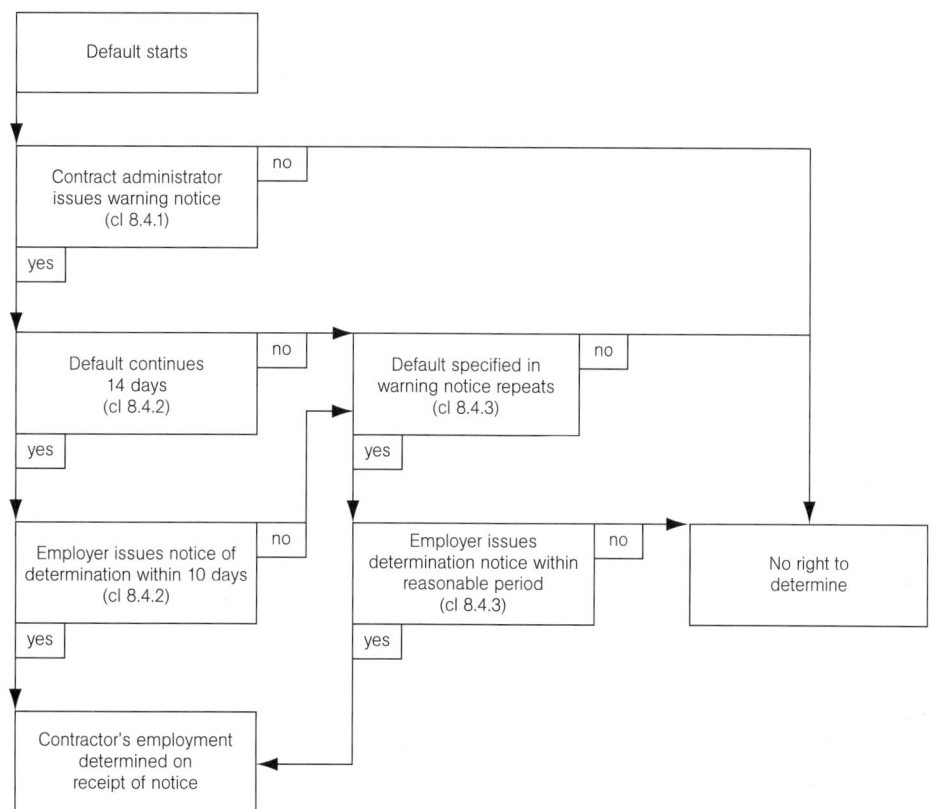

warning notice. In the case of *London Borough of Hounslow* v *Twickenham Garden Developments*, for example, the contract administrator's notice was heavily attacked by the defendants. In a more recent case, however, the contract administrator was found negligent because he failed to issue a notice (*West Faulkner Associates* v *London Borough of Newham*). It should be remembered that without the first 'warning notice' issued by the contract administrator the employer cannot issue the termination notice.

London Borough of Hounslow v Twickenham Garden Developments (1970) 7 BLR 81

The London Borough of Hounslow entered into a contract with Twickenham Garden Developments to carry out sub-structure works at Heston and Isleworth in Middlesex. The contract was on JCT63. Work on the contract stopped for approximately eight months due to a strike. After work resumed, the contract administrators issued a notice of default stating that the contractor had failed to proceed regularly and diligently and that unless there was an appreciable improvement the contract would be determined. The employers then proceeded to determine the contractor's employment. The contractor disputed the validity of the notices and the determination, and refused to stop work and leave the site. The Borough applied to the court for an injunction to remove the contractor. The judge emphasised that an injunction was a serious remedy and that before he could grant one there had to be clear and indisputable evidence of the merits of their case. The evidence put before him, which showed a significant drop in the amounts of monthly certificates and numbers of workmen on site, failed to provide this.

West Faulkner Associates v London Borough of Newham (1992) 61 BLR 81

West Faulkner were contract administrators engaged by the Borough for the refurbishment of a housing estate consisting of several blocks of flats. The residents of the estate were evacuated from their flats in stages to make way for the contractors, Moss, whom it had been agreed would carry out the work according to a programme of phased possession and completion, with each block to take nine weeks. Moss fell behind the programme almost immediately. However, they had a large workforce on the site and continually promised to revise their programme and working methods to address the problems of lateness, poor quality work and unsafe working practices that were drawn to their attention on numerous occasions by the contract administrators. In reality Moss remained completely disorganised, and there was no apparent improvement.

The contract administrators took the advice of quantity surveyors that the grounds of failing to proceed regularly and diligently would be difficult to prove, and decided not to issue a notice. As a consequence the Borough were unable to issue a notice of termination, had to negotiate a settlement with the contractors and dismissed the contract administrators, who then brought a claim for their fees. The judge decided that the contract administrators were in breach of contract in failing to give proper consideration to the use of the termination provisions. In his judgment he stated that:

> 'regularly and diligently' should be construed together and in essence they mean simply that the contractors must go about their work in such a way as to achieve their contractual obligations. This requires them to plan their work, to lead and manage their workforce, to provide sufficient and proper materials and to employ competent tradesmen, so that the Works are carried out to an acceptable standard and that all time, sequence and other provisions are fulfilled. (Judge Newey at page 139)

Insolvency of the contractor

10.11   Insolvency is the inability to pay debts as they become due for payment. Insolvent individuals may be declared bankrupt. Insolvent companies may be dealt with in a number of ways depending upon the circumstances, for example by voluntary liquidation (in which the company resolves to wind itself up); compulsory liquidation (under which the company is wound up by a court order); administrative receivership (a procedure to assist the rescue of a company under appointed receivers); an administration order (a court order given in response to a petition, again with the aim of rescue rather than liquidation, and managed by an appointed receiver); or voluntary arrangement (in which the company agrees terms with creditors over payment of debts). Procedures for dealing with insolvency are mainly subject to the Insolvency Act 1986 and the Insolvency Rules. Under these the person authorised to oversee statutory insolvency procedures is termed an insolvency practitioner.

10.12   Under SBC05 the contractor must notify the employer in writing in the event of liquidation or insolvency (cl 8·5·2). The employer is given an option to terminate (cl 8·5·1), or to consider a more constructive approach. This is to allow the appointed insolvency practitioner time to come up with a rescue package if possible. It is usually in the employer's interest to have the Works completed with as little additional delay and cost as possible, and a breathing space might allow all possibilities to be explored. During this period the employer is under no obligation to make further payment (cl 8·5·3·1), and the contractor's obligation to 'carry out and complete the Works and the design of the Contractor's Designed Portion' is suspended (cl 8·5·3·2). The employer can then either make an agreement to arrange for the work to continue, or terminate the employment of the contractor.

10.13   There are basically three options. The first is that arrangements could be made for the contractor to continue and complete the Works. Unless the insolvency practitioner has been able to arrange resource backing, this may not be a realistic option. If practical completion is near, however, and there is money due to the contractor, it can be advantageous to allow completion under the insolvency practitioner's control.

10.14   As an alternative, another contractor may be novated to complete the Works. On a 'true novation', the substitute contractor would take over all the original obligations and benefits (including completion to time and within contract sum). More likely is the third option which is 'conditional novation' whereby the contract completion date, etc. would be subject to re-negotiation, and the substitute contractor would probably want to disclaim liability for that part of the work undertaken by the original contractor.

10.15   Deciding on which of the options would best serve the interests of all the parties is a matter between the employer, no doubt advised by the contract administrator, and the insolvency practitioner. There might be merit in adopting one particular course of action, or there might be advantages in taking a more pragmatic approach. For example it may prove expeditious to continue initially with the original contractor under an interim arrangement until such time as novation can be arranged, or a completion contract negotiated.

Consequences of termination

10.16   If the employer exercises its right to terminate under clause 8·4, 8·5 or 8·6, then the only way to achieve completion will be by a new contractor of the employer's choice. The contract gives the employer the right to employ others under clause 8·7·1, to both carry out the Works and complete the Designed Portion. A completion contract might result from negotiation or competitive tender. The employer will have the right to:

- use any temporary buildings, plant etc. on site which are not owned by the original contractor, subject to the consent of the owner (cl 8·7·1);

- require the original contractor to assign the benefit of any sub-contracts to the employer (except if insolvent or in liquidation, and to the extent that the benefit is assignable) (cl 8·7·2·3).

10.17   The original contractor must remove temporary plant, etc. when required (including arranging for plant, etc. not owned by the contractor to be removed by the owner) (cl 8·7·2·1). The contractor must also provide the employer with two copies of all Design Documents, whether or not already provided (cl 8·7·2·2).

10.18   Following termination, the normal provisions regarding payment cease to apply (cl 8·7·3), and the employer would not be obliged to pay any outstanding certificates. (It should be noted, however, that the employer may still be obliged to pay amounts awarded by an adjudicator, see *Ferson* v *Levolux*.) A notional final account must be set out stating what is owed or owing, either in a statement prepared by the employer or in a contract administrator's certificate (cl 8·7·4). This must be done 'within a reasonable time after the completion of the Works', which allows the employer a period to assess its losses due to the termination.

10.19   One of the consequences of termination is that it often takes time for the contractor to effect an orderly withdrawal from site, and for the employer to establish the amounts outstanding before final payment. Should the employer decide not to continue with the construction of the Works after termination, the employer is required to notify the contractor in writing within six months of that notice (cl 8·8). Within a reasonable time of the notification (or within six months of termination, if no work is carried out and no notice issued) the employer must send the contractor a statement of the value of the works and losses suffered as required under clause 8·8.

Ferson Contractors Ltd v Levolux A. T. Ltd [2003] BLR 118

Ferson were contractors and Levolux sub-contractors on a GC/Wks sub-contract. A dispute arose regarding Levolux's second application for payment; £56,413 was claimed but only £4,753 paid. A withholding notice was issued which specified the amount but not the reason for withholding it. Levolux brought a claim to adjudication, and the adjudicator decided that the notice did not comply with section 111 of the Act, and that Ferson should pay the whole amount. Ferson refused to pay and Levolux sought enforcement of the decision. Prior to the adjudication Levolux had suspended work and Ferson, maintaining that the suspension was unlawful, had determined the contract. They now maintained that due to clause 29 which stated 'all sums of money that may be due or accruing due from the contractor's side to the sub-contractors shall cease to be due or accrue due' meant that they did not have to pay this amount. The CA upheld the judge of first instance decision that the amount should be paid: 'The contract must be construed so as to give effect to the intent of Parliament'.

Termination by the contractor

10.20   The contractor is given a reciprocal right to terminate its own employment under clause 8·9·1 in the event of specified defaults of the employer or specified suspension events under clause 8·9·2, or insolvency of the employer (cl 8·10). The specified events must have resulted in the suspension of the whole of the uncompleted Works for the continuous period stated in the Contract Particulars (if no period is entered, the default period is two months). In the case of specified defaults or suspension events a notice is required, which must specify the default or event. If the default or event continues for 14 days from receipt of the notice, the contractor may terminate the employment by a further notice up to 10 days from the expiry of the 14 days. Alternatively, if the employer ends the default or the suspension event ceases, and the contractor gives no further notice, should the employer repeat the default the contractor may terminate 'within a reasonable time after such repetition' (cl 8·9·3). These notices must be given by actual delivery, special delivery or recorded delivery, and not by fax or e-mail.

10.21   The grounds differ from those that give the employer the right to terminate. They include failure to pay an amount 'properly due' on a certificate, and obstruction of the issue of a certificate. There are also matters which relate directly to the duties of the contract administrator, where for example the carrying out of the whole or substantially the whole of the Works is suspended for a period stated in the Contract Particulars, due to contract administrator instructions under clauses 2·15, 3·14 or 3·15 (cl 8·9·2·1), or due to 'any impediment act or default' of the employer (cl 8·9·2·2). The period of suspension is a month unless some other period is stated. In long and complex projects a month may not be sufficient should unexpected technical problems arise, therefore it may be advisable to consider inserting a longer period.

10.22   Termination by the contractor is optional in the case of the employer's bankruptcy or insolvency (cl 8·10·1). The contractor must issue a notice and termination would take effect from the receipt of the notice.

10.23   The contractor must then remove from the site all temporary buildings, tools, etc. (cl 8·12·2). The contractor then prepares an account setting out the total value of the work at the date of termination, plus other costs relating to the termination as set out in clause 8·12·3. These may include such items as the cost of removal and any direct loss and/or damage consequent upon termination (cl 8·12·3·5). The contractor is in effect indemnified against any damages that may be caused as a result of the termination. This would not necessarily be the case if the contractor did not comply with the contractual provisions; in that case it might constitute repudiation.

Termination by either the employer or the contractor

10.24   Either party is given the right to terminate if the carrying out of the Works is wholly or substantially suspended for the continuous period stated in the Contract Particulars due to one or more of the events listed in clause 8·11·1 (if no period is entered, the default period is two months). These include loss or damage to the Works caused by the Specified Perils, and also contract administrator's instructions issued as a result of negligence or default of a local authority or statutory undertaker executing work solely in pursuance of its statutory obligations. The right of the contractor to terminate in the event of a Specified Peril is limited by the proviso that the event must not have been caused by the contractor's negligence (cl 8·11·2).

10.25   Notice may be given by either party and the employment of the contractor will be terminated seven days after receipt of the notice, unless the suspension is terminated within that period (cl 8·11·1).

10.26   Detailed provisions are set out regarding the consequences of the termination. The contractor must remove all temporary buildings, tools etc. from the site. Within two months of termination the contractor must provide the employer with all the necessary documents for the purposes of preparing an account or, if the employer prefers, the contractor may prepare the account. The items that are to be considered for inclusion in the account are as before set out in clause 8·12·3. If termination is due to an event caused by a Specified Peril, which was in turn caused by the negligence or default of the employer, then the contractor may be due an item for direct loss and/or damage suffered as a result of the termination but otherwise these losses will not be due (cl 8·12·14).

11 Dispute resolution

11.1   SBC05 refers to four methods of dispute resolution: mediation; adjudication; arbitration; and legal proceedings. One of these, adjudication, is a statutory right, and if one party wishes to use this method the other has no alternative but to concur. Mediation is a voluntary process and therefore requires the agreement of both parties at the time of the dispute if it is to take place. If neither of these processes is used, or if either party is dissatisfied with the decision of an adjudicator, then the dispute will have to be terminated by arbitration or litigation. SBC05 requires the parties to decide in advance which of these two processes will be used.

11.2   There are therefore stages, either before or during the contract, where the parties have the opportunity to agree a preferred course of action. It is important for the contract administrator to understand and to be able to advise on these methods. The dispute may have arisen out of a decision made by the contract administrator, and the contract administrator may find that he or she becomes involved. The contract administrator may be asked to give evidence, so a basic understanding of the procedures involved is essential.

11.3   Before any of the more formal procedures are initiated, there may be a period of negotiation where the parties attempt to resolve their differences themselves. This might be the best solution to the problem, but the contract administrator should tread carefully before becoming involved in such negotiations. The contract administrator may be of great assistance in advising the client and providing information, but has no authority to negotiate amendments to the terms of the contract or make ad-hoc agreements on behalf of the client. Even if the client gives the contract administrator an extended authority to negotiate a settlement, where the dispute involves complex legal points, a lawyer may best handle the negotiations.

Alternative dispute resolution

11.4   If negotiations fail to achieve an agreement, the parties may submit the dispute to 'Alternative Dispute Resolution' (ADR), a name used to cover methods such as conciliation, mediation and the mini-trial. SBC05 draws attention to the possibility of using mediation in clause 9·1, which refers to the JCT Guide (SBC/G). The Guide does not itself set out or advocate any particular procedure to be used in mediation, stating that 'such choices are frequently better made by the Parties when the dispute has actually arisen'. The parties could, of course, supplement SBC05 by selecting a procedure or mediator appointing body, and setting this out in the terms of their appointment. As mediation is a consensual process, any individual reference to mediation would have to be supported by both parties.

11.5   Usually a mediator is appointed jointly by the parties and will meet with the parties together and separately in an attempt to resolve the differences. The outcome is often in the form of a recommendation which, if acceptable, can be signed as a legally binding agreement.

This would then be enforceable in the same way as any other contract. However, if the recommendation is not acceptable to one of the parties and is not signed as a binding agreement, it cannot be imposed by law, and so the time spent on the mediation may appear to have been wasted.

11.6   Nevertheless, there can be many advantages to mediation. Unlike adjudication, arbitration or litigation, it is a non-adversarial process which tends to forge good relationships between the parties. Imposed solutions may leave at least one of the parties dissatisfied and may make it very difficult to work together in the future. If the parties are keen to promote a long-term business relationship they should give mediation serious consideration. Even if mediation does not result in a complete solution it has been found in practice that it can help to clear the air on some of the issues involved and establishes common ground. This, in turn, might then pave the way for shorter and possibly less acrimonious arbitration or litigation.

Adjudication

11.7   The Housing Grants, Construction and Regeneration Act 1996 requires that parties to the construction contracts falling within the definition set out in the Act have the right to refer any dispute to a process of adjudication which complies with requirements stipulated in the Act. Article 7 of SBC05 re-states this right, and refers to clause 9·2, which states that where a party decides to exercise this right 'the Scheme shall apply'. This is referring to the Scheme for Construction Contracts, a piece of secondary legislation which sets out a procedure for the appointment of the adjudicator and the conduct of the adjudication. The Scheme takes effect as implied terms in a contract, if and to the extent that the parties have failed to agree on a procedure that complies with the Act.

11.8   By stating 'the Scheme shall apply', SBC05 is effectively annexing the provisions of the Scheme to the form, which therefore become a binding part of the agreement between the parties. Clause 9·2, however, makes its application subject to certain conditions which relate to the appointment of the adjudicator.

11.9   Under SBC05 the adjudicator may either be named in the Contract Particulars, or nominated by the nominating body identified in the Contract Particulars. A named adjudicator will normally enter into the JCT Adjudication Agreement Named Adjudicator (Adj/N) with the parties at the time the main contract is entered into.

11.10   The party wishing to refer a dispute to adjudication must first give notice under section 1(1) of the Scheme (see Figure 24). The notice should identify briefly the dispute or difference, give details of where and when it has arisen, set out the nature of the redress sought, and include the names and addresses of the parties, including any specified for the giving of notices (s.1(3)). If no adjudicator is named, the parties may either agree an adjudicator or either party may apply to the 'nominator' identified in the Contract Particulars (s.2(1)). If no nominator has been selected, then the contract states that the referring party may apply to any of the nominators listed in the Contract Particulars. The adjudicator will then send terms of appointment to the parties. In addition to the form (or a named

Figure 24 Appointment of the adjudicator

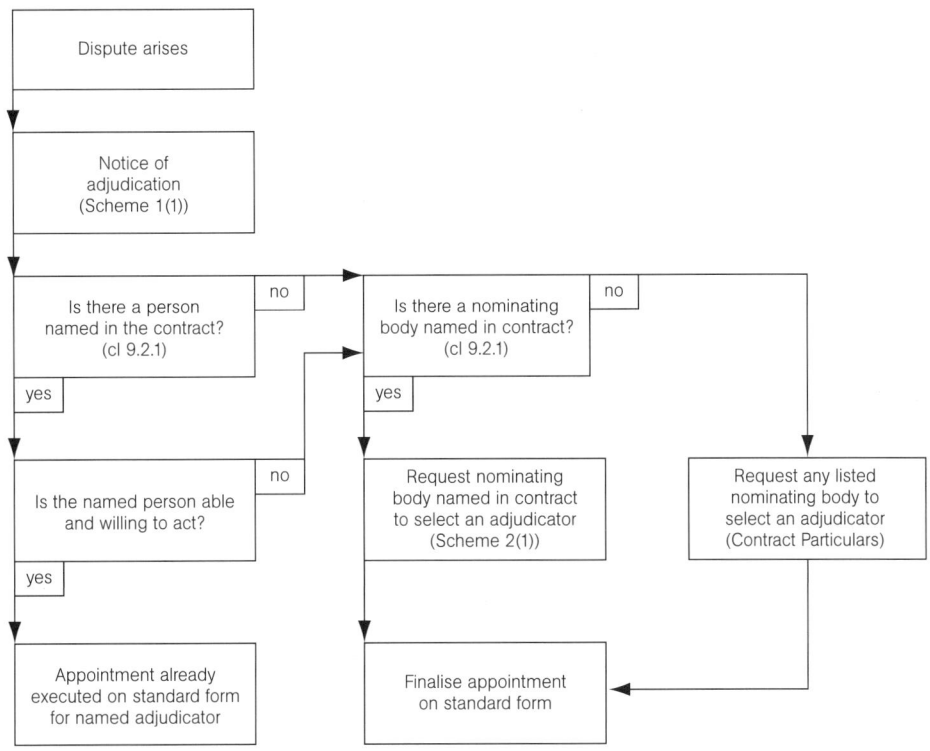

adjudicator, the JCT also publishes an adjudication agreement (Adj) for use in this situation.

11.11 The Scheme does not stipulate any qualifications in order to be an adjudicator, but does state that the adjudicator 'should be a natural person acting in his personal capacity' and should not be an employee of either of the parties (s.4). In addition, SBC05 requires that where the dispute relates to clause 3.18.4 (repeat testing), that the person appointed shall, 'where practicable' be 'an individual with appropriate expertise in the specialist area or discipline relevant to the instruction or issue in dispute' (cl 9.2.2). Where the person does not have the appropriate expertise he or she must appoint an independent expert to advise and report.

11.12 The adjudicator is required to act impartially must avoid unnecessary expense (s.12), and is not liable for anything done or omitted when acting properly as an adjudicator (s.26).

11.13 The referring party must refer the dispute to the selected adjudicator within seven days of the date of the notice (s.7(1)). The referral will normally include particulars of the dispute, and must include a copy of, or relevant extracts from, the contract, and any material it wishes the adjudicator to consider (s.7(2)). A copy of the referral must be sent to the other party (s.7(3)).

11.14   The adjudicator will then set out the procedure to be followed. A preliminary meeting may be held to discuss this, otherwise the adjudicator may send the procedure and timetable to both parties. The party who did not initiate the adjudication (the responding party) will be required to respond by a stipulated deadline. The adjudicator is likely to hold a short hearing of a few days at which the parties can put forward further arguments and evidence. There may also be a site visit. Occasionally it may be possible to do the whole thing by correspondence (often termed 'documents only').

11.15   The adjudicator is given considerable powers under the scheme, including the right to take the initiative in obtaining the facts and the law, the right to issue directions, the right to revise decisions and certificates of the contract administrator, the right to carry out tests (subject to obtaining necessary consents), and the right to obtain from others necessary information and advice. The adjudicator must give advance notice if intending to take legal or technical advice.

11.16   The Act requires that the decision is reached within 28 days of referral, but it does not state how this date is to be established (section 108(2)(c)). Under the Scheme the 28 days starts to run from the date of the referral notice, which may be earlier than the receipt of all the information by the adjudicator (s.19(1)). The period can be extended by up to 14 days by the referring party, and further by agreement between the parties. The decision must be delivered forthwith to the parties, and the adjudicator may not retain it pending payment of the fee. The provisions state that the adjudicator must give reasons for the decision if requested to do so by the parties (s.22).

11.17   The parties must meet their own costs of the adjudication, unless they have agreed that the adjudicator shall have the power to award costs. The adjudicator is entitled to charge fees, and expenses that are 'reasonably incurred'. The adjudicator is entitled to apportion these between the parties, but if not the parties will bear these in equal proportions. The parties are liable jointly and severally for the adjudicator's fees and expenses, which means that in the event of default by one party, the other party becomes liable for the full amount.

11.18   The adjudicator's decision will be final and binding on the parties 'until the dispute is finally determined by legal proceedings, by arbitration, or by agreement between the parties'. The effect of this is that if either party is dissatisfied with the decision, they may raise the dispute again in arbitration or litigation as indicated in the Contract Particulars, or negotiate a fresh agreement with the other party. In all cases, however, they remain bound by the decision and must comply with it until the final outcome is determined.

11.19   If either party refuses to comply with the decision, the other may seek to enforce it through the courts. Generally actions regarding adjudicators' decisions have been dealt with promptly by the courts and the recalcitrant party has been required to comply.

Arbitration

11.20 Arbitration refers to proceedings in which the arbitrator has power derived from a written agreement between the parties to a contract, and which is subject to the provisions of the Arbitration Act 1996. Arbitration awards are enforceable at law. An arbitrator's award can be subject to appeal on limited grounds.

11.21 If arbitration is preferred to litigation as the method for final determination of disputes, then this is confirmed by indicating in the Contract Particulars that Article 8 will apply (note that if no entry is made, the default process will be litigation). The arbitration provisions are set out in clause 9·3, which refers to the Construction Industry Model Arbitration Rules (the Rules). The Arbitration Act 1996 confers wide powers on the arbitrator unless the parties have agreed otherwise, but leaves detailed procedural matters to be agreed between the parties or, if not so agreed, to be decided by the arbitrator. To avoid problems arising, it is advisable to agree as much as possible of the procedural matters in advance, and SBC05 does this by incorporating the Rules, which are very clearly written and self-explanatory. The specific edition referred to is the 2005 edition published by the JCT, which amends some of the Rules and incorporates supplementary and advisory procedures.

11.22 The party wishing to refer the dispute to arbitration must give notice as required by SBC05 clause 9·4 and Rule 2.1, identifying briefly the dispute and requiring the party to agree to the appointment of an arbitrator. If the parties fail to agree within 14 days, either party may apply to the 'appointor', selected in advance from a list of organisations set out in the Contract Particulars. If no appointor is selected, then the contract states that the appointor will be the RIBA. Under Rule 2.5 the arbitrator's appointment takes effect when he or she agrees to act, and is not subject to first reaching agreement with the parties on matters such as fees.

11.23 The arbitrator has the right and the duty to decide all procedural matters, subject to the parties' right to agree any matter (Rule 5.1). Within 14 days of appointment the parties must each send the arbitrator and each other a note indicating the nature of the dispute and amounts in issue, the estimated length for the hearing, if necessary, and the procedures to be followed (Rule 6.2, as amended). The arbitrator must hold a preliminary meeting within 21 days of appointment to discuss these matters (Rule 6.3 as amended). The first decision to make is whether Rule 7 (short hearing), Rule 8 (documents only) or Rule 9 (full procedure) is to apply. The decision will depend on the scale and type of dispute.

11.24 Under all three Rules referred to above, the parties exchange statements of claim and of defence, together with copies of documents and witness statements on which they intend to rely. Under Rule 8 the arbitrator makes his or her award based on the documentary evidence only. Under Rule 9 the arbitrator will hold a hearing at which the parties or their representatives can put forward further arguments and evidence. There may also be a site visit. The JCT amendments set out time limits for these procedures.

11.25 Under Rule 7 (as amended) a hearing is to be held within 21 days of the date when Rule 7 becomes applicable, and the parties must exchange documents not later than seven days prior to the hearing. The hearing should be not more than one day. The arbitrator publishes the award within one month of the hearing. The parties bear their own costs.

11.26   The arbitrator is given a wide range of powers under Rule 4, including the power to obtain advice (Rule 4.2), the powers set out in section 38 of the Arbitration Act 1996 (Rule 4.3), the power to order the preservation of work, goods and materials even though they are a part of work that is continuing (Rule 4.4), the power to request the parties to carry out tests (Rule 4.5), and the power to award security for costs. Under clause 9·5 of SBC05 the arbitrator is also given wide powers to review and revise any certificate, opinion, decision, requirement or notice and to disregard them if need be, where seeking to determine all matters in dispute.

11.27   Costs are normally awarded on a judicial basis, i.e. the loser will pay the winner's costs (Rule 13.1). The arbitrator will be entitled to charge fees and expenses and will apportion those fees between the parties on the same basis. The parties are jointly and severally liable to the arbitrator for fees and expenses incurred.

Arbitration and adjudication

11.28   Under article 8 any dispute that has been referred to an adjudicator may be referred to arbitration if either party requires this. Clause 1·10·4 states that even where the decision has been given after the Final Certificate is issued, either party may refer the dispute to arbitration, provided the arbitration is commenced within 28 days of the adjudicator's decision. It is not entirely clear, however, whether this will affect the date at which the Final Certificate will become conclusive evidence of the matters listed under clause 1·10·1, which may of course have been the matters disputed under the adjudication.

Arbitration or litigation

11.29   As stated in the introduction to this section, SBC05 contains alternative provisions for arbitration and litigation in articles 8 and 9, and a choice has to be made before tender documents are sent out. Both processes give rise to binding and enforceable decisions. Both tend to be lengthy and expensive, although there are provisions for short forms of arbitration.

11.30   Litigation cases involving claims for amounts greater than £25,000 are normally heard in the High Court, and construction cases are usually heard in the Technology and Construction Court, a specialist department of the High Court which deals with technical or scientific cases. Procedures in court follow the Rules of the Supreme Court, with the timetable and other detailed arrangements being determined by the court. A judge will hear the case, and in the High Court a barrister must represent the parties.

11.31   Disputes in building contracts have traditionally been settled by arbitration. Arbitrators are usually senior and experienced members of one of the construction professions, and for many years it was felt that they had a greater understanding of construction projects and the disputes that arise than might be found in the courts. These days, however, the judges of the Technology and Construction Court have extensive experience of technical construction disputes. The high standards now evident in these courts are likely to be matched in practice by only a few arbitrators.

11.32 The court has powers to order that actions regarding related matters are joined (for example where disputes between an employer and contractor, and contractor and nominated sub-contractor, concern the same issues). This is much more difficult to achieve in arbitration. Even if all parties have agreed to the CIMA Rules, the appointing bodies must have been alerted and agreed to appoint the same arbitrator (Rules 2.6 and 2.7). If the same arbitrator is appointed, he or she may order concurrent hearings (Rule 3.7) but may only order consolidated proceedings with all the parties' consent (Rule 3.9), which is often difficult to obtain. The court's powers may therefore be an advantage in multi-party disputes, to avoid duplication of hearings and possible conflicting outcomes.

11.33 There remain, however, two key advantages to using arbitration. The first is that in arbitration the proceedings can be kept private – this is something which is usually of paramount importance to construction professionals and companies, and is often a deciding factor in selecting arbitration. In court, the proceedings are open to the public and the press, and the judgment is published and widely available.

11.34 The second advantage to the parties is that the arbitration process is consensual. The parties are free to agree on timing, place, representation and the individual arbitrator. This autonomy carries with it the benefits of increased convenience, and possibly savings in time and expense. The parties avoid long waiting lists currently running at the High Court, and choose a convenient time and place for the hearing. In arbitration, however, the parties have to pay the arbitrator and for the cost of renting premises in which the hearing is held.

11.35 It should perhaps be noted that even where parties have selected arbitration under article 8, it is still open for them to elect litigation once a dispute develops. If, however, one party commences court proceedings, the other may ask the court to stay the proceedings on the grounds that an arbitration agreement already exists (except for actions to enforce an adjudicator's decision, as article 8 excludes such disputes from the jurisdiction of the arbitrator). If, on the other hand, the parties had originally selected litigation, this would not prevent them from subsequently agreeing to take a dispute to arbitration, but in such cases they would also need to agree which procedural rules are to apply.

Appendix A: *Clause comparison table: Destinations*

Table of destinations

	JCT98 (1-4)	SBC05	subject	change
Recitals	first	first	work	footnote changed from 'nature' to 'nature and location'
	second	second	Bills/Activity Schedule	Activity Schedule now explained in footnote 3
	third	third	drawings	drawings may be initialled, footnote added
	fourth	fourth	tax status	amended to refer to the Construction Industry Scheme, footnote added
	fifth		CDM	omitted
	sixth	fifth	information release schedule	unchanged
	seventh		bond	omitted
	second (CDPS/SC)	seventh	design portion	equivalent to 2nd recital in CDP/SC supplement
	third (CDPS/SC)	eighth	design portion	part of 3rd recital in CDP/SC supplement
	fourth (CDPS/SC)	ninth	design portion	equivalent to 4th recital in CDP/SC supplement
	tenth (CDPS/SC)	tenth	design portion	equivalent to 10th recital in CDP/SC supplement
Articles	Article 1	Article 1	contractor's obligations	wording simplified
	Article 2	Article 2	Contract Sum	wording simplified
	Article 3	Article 3	architect/contract administrator	wording simplified, employer's re-nomination obligation transferred to clause 3·5
	Article 4	Article 4	quantity surveyor	wording simplified, employer's re-nomination obligation transferred to clause 3·5
	Article 5	Article 7	adjudication	unchanged
	Article 6.1	Article 5	planning supervisor	wording simplified
	Article 6.2	Article 6	principal contractor	now allows for someone other than the contractor to act as principal contractor

Table of destinations Continued

Article	JCT98 (1-4)	SBC05	subject	change
	Article 7A	Article 8	arbitration	Arbitration no longer the default
	Article 7B	Article 9	legal proceedings	new – jurisdictional – English courts to have jurisdiction
	Interpretation			
clauses	1·1	1·2	reference to clauses etc.	now also covers references to and within Schedules
	1·2	1·3	agreement etc. to be read as a whole	wording simplified, amalgamated with 2·2·1
	1·3	1·1, 6·8	definitions	new defined terms and changes to existing
	1·4		*not used*	
	1·5	3·6	Contractor's responsibility	unchanged
	1·6	3·26	CDM re-appointments	minor changes
	1·7	1·7	serving notices etc.	re-structured and clarified, an address may be entered in the Contract Particulars
	1·8	1·5	reckoning periods of days	unchanged
	1·9	3·3	employer's representative	re-drafted, power to terminate appointment by notice added
	1·10	1·12	applicable law	wording simplified
	1·11	1·8	electronic communications	EDI omitted, parties may agree their own procedures
	1·12	1·6	third party rights	reference to granting Third Party Rights under 7A and 7B included
	Contractor's obligations			
	2·1	2·1, 2·3·3	general obligations	rewritten: 'where and to the extent that approval….' etc. removed, goes to clause 2·3·3, 'workmanlike manner' etc. included from clause 8·1·3, compliance with statutory requirements from 6·1·1
	2·1·1		*SC amendment*	omitted
	2·1·2	2·2	*CDP amendments*	re-structured but essentially the same
	2·1·3	2·3·3	*CDP amendments*	unchanged
	2·2·1	1·3	hierarchy	wording simplified
	2·2·2·1	2·13·1	preparation of bills	unchanged
	2·2·2·2	2·14·1, 2·14·3	errors in bills	essentially the same, references to performance specified work removed
	2·2·2·3	2·14·4	errors in Contractor's Proposals	unchanged

Table of destinations Continued

clauses	JCT98 (1-4)	SBC05	subject	change
	2·3	2·15	discrepancies	contractor to give written notice of error etc. 'with appropriate details'
	2·4		discrepancies	removed (performance specified work)
	2·5·1	2·16·1	discrepancies, CDP documents	unchanged
	2·6·1	2·9·2	*CDP info supply*	unchanged
	2·6·2	2·9·3	*CDP info supply*	reference to new Contractor's Design Submission Procedure added
	2·7·1	2·19·1	*CDP design liability*	unchanged
	2·7·2	2·19·2	*CDP design liability*	second part of clause removed
	2·7·3	2·19·3	*CDP design liability*	minor changes
	2·7·4			omitted
	2·8	3·10·3	*effect on CDPS*	unchanged
	2·10	2·20	*errors, CDP*	minor changes
	Contract sum			
	3	4·4	contract sum	minor changes
	Architect's instructions			
	4·1·1	3·10·1, 3·10·2	compliance	minor changes
	4·1·2	3·11	failure to comply	minor changes
	4·2	3·13	empowerment	minor changes
	4·3	3·12	instructions in writing etc.	wording simplified
	Contract Documents			
	5·1	2·8·1	contract document	changed to be in the custody of the employer
	5·2	2·8·2	copies to contractor	unchanged
	5·3·1·1	2·9·1·1	schedules	unchanged
	5·3·1·2	2·9·1·2	master programme	unchanged
	5·3·2	2·9·1	schedules	unchanged
	5·4·1	2·11	information release schedule	unchanged
	5·4·2	2·12·1	supply info	paragraph split into three, wording tidied up
	5·5	2·8·3	documents to be kept on site	unchanged

Table of destinations Continued

clauses	JCT98 (1-4)	SBC05	subject	change
	5·6			omitted
	5·7	2·8·4	use of documents	minor changes
	5·8	1·9	issue of certificates	unchanged
	5·9		performance spec work	omitted
	5·10	2·40	*CDP as-built drawings*	wording simplified
	Statutory obligations			
	6·1·1	2·1		omitted, general obligation included under clause 2·1
	6·1·2	2·17·1	divergences from statutory requirements	both parties must now notify the other
	6·1·3	2·17·2	instructions re above	re-structured
	6·1·4	2·18	emergency compliance	minor changes
	6·1·5	2·17·3	contractor not liable	minor changes
	6·1·6			*omitted*
	6·1·7			*omitted*
	6·1·8	2·17·1	divergence, CDP docs	amalgamated with 6·1·2
	6·1·9	2·18·1	emergency compliance	amalgamated with 6·1·4
	6·2	2·21	statutory fees	re-structured
	6·3	3·7·3	statutory undertakers	minor changes
	6A·1	3·25·1	employer CDM duties	note new 'catch all' introductory statement, otherwise unchanged
	6A·2	3·25·2	contractor CDM duties	reference to 15(4) CDM regs removed and specific obligation included
	6A·3	3·26	new Principal Contractor	minor changes
	6A·4	3·25·3	health and safety file	reference to 14 CDM regs removed and specific obligation included
	Levels, setting out			
	7	2·10	setting out	'at ground level' removed
	Work, materials, goods			
	8·1·1	2·3·1, 2·3·3	materials and goods	re-written, reference to performance specified work removed, 'so far as procurable' remains, substitution of goods etc. added, reference to 'reasonable satisfaction' goes to 2·3·3

Table of destinations Continued

clauses	JCT98 (1–4)	SBC05	subject	change
	8·1·2	2·3·2, 2·3·3	workmanship	re-written, reference to performance specified work removed, reference to 'a standard appropriate to the works' goes to 2·3·3, reference to 'reasonable satisfaction' goes to 2·3·3
	8·1·3	2·1	workmanlike manner	included under 2·1
	8·1·4	2·3·1	substitution	minor changes
	8·2·1	2·3·4	vouchers	'vouchers' replaced with 'reasonable proof'
	8·2·2	3·20	executed work	new requirement to give reasons for dissatisfaction
	8·3	3·17	inspection/tests	wording simplified
	8·4·1	3·18·1	defective work	wording simplified
	8·4·2	3·18·2	defective work	wording simplified
	8·4·3	3·18·3	defective work	minor changes
	8·4·4	3·18·4	defective work	minor changes
	8·5	3·19	workmanship	wording simplified
	8·6	3·21	persons	unchanged
Royalties				
	9·1	2·22	royalties	unchanged
	9·2	2·23	royalties	minor changes
Person-in-charge				
	10	3·2	person-in-charge	unchanged
Access for Architect to the works				
	11	3·1	access to the works	references to NSC removed
Clerk of works				
	12	3·4	clerk of works	unchanged
Variations and provisional sums				
	13·1·1	5·1·1	alteration to the works	unchanged
	13·1·2	5·1·2	restrictions, etc.	unchanged
	13·1·3		NSCs	omitted
	13·2	3·14	variation instructions	unchanged, except that 13·2·3 is moved to clause 5·3·1 and replaced by part of 13·2·1 referring to variations to employer's requirements
	13·2·3	5·3·1, 5·3·2	13A quotations	moved to clause 5·3 on valuing variations

Table of destinations Continued

clauses	JCT98 (1-4)	SBC05	subject	change
	13·3·1	3·16	AI provisional sums	unchanged
	13·3·2		AI sub-contracts	omitted
	13·4·1·1	5·2·1, 5·2·2	valuation procedure	simplified to say either agreed or by QS (Alternative A and 13A omitted), new reference to 'Schedule 2 Quotation'
	13·4·1·2		Alternative A	omitted
	13·4·1·3	5·2·1	Alternative B	inorporated in clause 5·2·1
	13·4·2		prime cost work	omitted
	13·5·1	5·6·1	valuation rules	unchanged
	13·5·1·1	5·6·1·1	measurable work	unchanged
	13·5·1·2	5·6·1·2	measurable work	unchanged
	13·5·1·3	5·6·1·3	measurable work	unchanged
	13·5·1·4	5·6·1·3	measurable work	unchanged
	13·5·1·5	5·6·1·4	measurable work	unchanged
	13·5·2	5·6·2	omissions	unchanged
	13·5·3	5·6·3	measurement	unchanged
	13·5·4	5·7	daywork	reference to Building Employers Confederation removed
	13·5·5	5·9	change of conditions	clarified
	13·5·6		performance specified work	omitted
	13·5·7	5·10	other work	unchanged
	13·6·1	5·4	contractor's right to be present	unchanged
	13·6·2			see above
	13·7	5·5	giving effect	simplified to reflect new valuation alternatives
	13·8	5·8	variations CDP	minor changes to wording
	13A	5·3	contractor's quotation	13A procedures replaced by new Schedule 2 Quotations (except 13A·8 goes to 5·3·3): wording largely the same with some minor tidying up
	13A·1·1	5·3·1	info provided	only first reference to information retained
	13A·8	5·3·3	variation to 13A work	unchanged
Contract Sum				
	14·1	4·1	contract sum	unchanged
	14·2	4·2	adjustment	unchanged

Table of destinations Continued

clauses	JCT98 (1-4)	SBC05	subject	change
VAT				
15·1			VAT – definition	omitted
15·2	4·6·1		VAT	considerably simplified
15·3	4·6·2		VAT	minor changes
Materials and goods unfixed or off-site				
16·1	2·24		unfixed	unchanged
16·2	2·25		off site	unchanged
Practical completion and defects liability				
17·1	2·30		Practical Completion Certificate	performance specified work removed
17·2	2·38		schedule defects/making good	re-structured, minor changes, reference to frost damage omitted
17·3	2·38·2		schedule defects/making good	re-structured, minor changes
17·4	2·39		certification	essentially the same, title changed to 'Certificate of Making Good'
17·5			frost	omitted
Partial possession				
18·1	2·33		statement	minor changes
18·1·1	2·34		PC date	minor changes
18·1·2	2·35		defects	minor changes
18·1·3	2·36		insurance	minor changes
18·1·4	2·37		LADs	wording simplified
Assignment and sub-contracts				
19·1·1	7·1		assignment	'subject to clause 7·2' introduced, ie, to rights of enforcement
19·1·2	7·2		right to enforce	minor changes
19·2·1			definition	omitted
19·2·2	3·7·1		consent	minor changes
19·2·3	3·7·2		CDP sub-let	minor changes
19·3·1	3·8·1		listed sub-contractors	minor changes
19·3·2	3·8·2, 3·8·3		listed sub-contractors	minor changes
19·3·3	3·8·4		listed sub-contractors	minor changes
19·4	3·9		sub-letting conditions	unchanged
19·4·1	3·9·1		termination sub-contract	unchanged

Table of destinations Continued

clauses	JCT98 (1-4)	SBC05	subject	change
	19·4·2·1	3·9·2·1	removal materials	minor changes
	19·4·2·2	3·9·2·1·1	property vesting	unchanged
	19·4·2·3	3·9·2·1·2	property vesting	minor changes
	19·4·2·4	3·9·2·5	vesting, listed items	minor changes, reference to 'off-site' changed to 'listed'
	19·4·3	3·9·2·3	interest	2nd half of clause omitted
	19·5·1		NSCs	omitted
	19·5·2		NSCs	omitted
Injury to persons or property				
	20·1	6·1	personal injury	shortened: last section replaced by term 'Employer's Persons'
	20·2	6·2	property	shortened: central section replaced by term 'Contractor's Persons', exception due to Employer/Employer's persons removed
	20·3·1	6·3·1	works excluded	re-structured
	20·3·2	6·3·3	partial possession	unchanged
Insurance against injury to persons or property				
	21·1·1	6·4·1	contractor's insurance	'person under a contract' replaced by 'any employee', re-structured
	21·1·2	6·4·2	proof	minor changes
	21·1·3	6·4·3	default	minor changes
	21·2·1	6·5·1	optional non-negligent damage	minor changes
	21·2·2	6·5·2		minor changes
	21·2·3	6·5·3	amounts added	unchanged
	21·2·4		default	omitted
	21·3	6·6	excepted risks	unchanged
Insurance of the Works				
	22·1	6·7	alternatives	re-written: options A, B, C now set out in Schedule 3, new terminology
	22·2	6·8	definitions	excepted risks and specified perils definitions now included, definition of terrorism changed
	22·3	6·9	recognition/waiver	wording simplified
	22A·1	S·3 A·1	insurance	minor changes, reference to VAT dropped note 15% default rate for professional fees
	22A·2	S·3 A·2	insurance	minor changes

Table of destinations Continued

clauses	JCT98 (1-4)	SBC05	subject	change
	22A·3	S·3 A·3	insurance	minor changes
	22A·4	S·3 A·4	insurance	minor changes
	22A·5·1	6·10·1	terrorism	minor changes
	22A·5·2	6·10·2, 6·10·3	terrorism	re-written
	22A·5·3	6·10·4	terrorism	minor changes
	22A·5·4	S·3 A·5	terrorism	minor changes
	22B·1	S·3 B·1	insurance	minor changes, reference to VAT dropped
	22B·2	S·3 B·2	insurance	minor changes, new B2·2 relating to a Local Authority
	22B·3	S·3 B·3	insurance	wording simplified
	22B·4·1	6·10·1	terrorism	minor changes
	22B·4·2	6·10·2, 6·10·3	terrorism	re-written
	22B·4·3	6·10·4	terrorism	minor changes
	22C·1	S·3 C·1	insurance	minor changes, reference to VAT dropped
	22C·1A	6·10·1–6·10·4	insurance	re-structured, minor changes
	22C·2	S·3 C·2	insurance	minor changes, reference to VAT dropped
	22C·3	S·3 C·3	insurance	minor changes, new C2·2 relating to a Local Authority
	22C·4	S·3 C·4	insurance	wording simplified
	22C·5·1	6·10·1	terrorism	minor changes
	22C·5·2	6·10·2, 6·10·3	terrorism	re-written
	22C·5·3	6·10·4	terrorism	minor changes
	22D		insurance LADs	omitted
	22FC·1	6·13	application	unchanged
	22FC·2	6·14	compliance	re-written – 2 paragraphs reduced to one sentence!
	22FC·3	6·15	breach of FC	minor rewording – term 'Remedial Date' removed
	22FC·5	6·16	amendments to FC	unchanged
	Date of possession			
	23·1·1	2·4	possession/ progress	minor changes
	23·1·2	2·5	deferment	minor changes
	23·2	3·15	postponement	clause removed and grouped with other clauses relating to instructions

Table of destinations Continued

clauses	JCT98 (1-4)	SBC05	subject	change
	23·3·1	2·4·1/2·4·2	retention of possession	amalgamated with 23·1·1
	23·3·2	2·6·1	use by employer	wording change to 'for storage or otherwise' ('of his goods' removed)
	23·3·3	2·6·2	use by employer	wording simplified
Damages for non-completion				
	24·1	2·31	non-completion cert	unchanged
	24·2·1	2·32·1, 2·32·2	notice of LADs	re-structured and simplified
	24·2·2	2·32·3	re-payment	unchanged
	24·2·3	2·32·4	new notice	expanded and clarified
Extensions of time				
	25·1	2·26	definitions	expanded
	25·2·1	2·27·1	notice	unchanged
	25·2·2	2·27·2	notice	reference to concurrent delay removed
	25·2·3	2·27·3	further notices	re-written, much clearer and firmer obligation
	25·3·1	2·28·1– 2·28·3	time limits	receipt of 'estimate' omitted, date to be fixed 'as soon as is reasonably practicable'. CA to 'endeavour' to notify by completion date
	25·3·1·1	2·28·1·1	fixing EOT	minor changes
	25·3·1·2	2·28·1·2	fixing EOT	minor changes
	25·3·1·3	2·28·3·1	fixing EOT	reworded, now has to state EOT for each event
	25·3·1·4	2·28·3·2	fixing EOT	re-written, CA now to state reductions due to omissions
	25·3·2	2·28·4, 2·28·6·4	reducing time	shortened, references to 13A quotations removed, goes to 2·28·6·4 – 'pre-agreed adjustment'
	25·3·3	2·28·5, 2·28·6	final review	largely unchanged, references to 13A quotations removed, goes to 2·28·6·4 – 'pre-agreed adjustment'
	25·3·4	2·28·6	provisos	2 of 4 provisos re-grouped from other clauses
	25·3·5		NSC	omitted
	25·3·6	2·28·6·3	proviso	unchanged
	25·4·1	2·29·8	force majeur	unchanged
	25·4·2	2·29·13	weather	unchanged

Table of destinations Continued

clauses	JCT98 (1-4)	SBC05	subject	change
	25·4·3	2·29·9	specified perils	unchanged
	25·4·4	2·29·10, 2·29·11	civil commotion strikes	terrorism added (25·4·16), strikes, etc. moved to separate event
	25·4·5·1	2·29·1, 2·29·2	variations, instructions	variations a separate event
	25·4·5·2	2·29·2	tests	minor changes
	25·4·6		information release	omitted
	25·4·7		NSC's	omitted
	25·4·8·1		employer's direct labour	omitted
	25·4·9	2·29·12	statutory powers	large section removed
	25·4·10		inability to get labour etc.	omitted
	25·4·11	2·29·7	statutory work	unchanged
	25·4·12		ingress/egress	omitted
	25·4·13	2·29·3	deferment	minor changes
	25·4·14	2·29·4	approx quantities	minor changes
	25·4·15		performance spec work	omitted
	25·4·16	2·29·10	terrorism	combined with civil comm
	25·4·17		CDM	omitted
	25·4·18	2·29·5	suspension	minor changes
	25·4·19	2·29·6	any impediment	qualification 'save as provided (above)' removed, now covers some events which were previously specifically listed
	Loss and expense			
	26·1	4·23	loss and expense	wording tidied up, refers to 'Relevant Matters'
	26·2	4·24	matters	now termed 'Relevant Matters'
	26·2·1		information release	omitted
	26·2·2	4·24·2·2	tests	phrase 'unless provided for in the Contract Bills' added
	26·2·3	4·24·2·3	discrepancies	minor changes
	26·2·4·1		employer's direct labour	omitted
	26·2·4·2		employer's direct labour	omitted
	26·2·5	4·24·2·1	postponement	
	26·2·6		ingress/egress	omitted
	26·2·7	4·24·1, 4·24·2·1	variations, provisional sums	re-written

Table of destinations Continued

clauses JCT98 (1-4)	SBC05	subject	change
26·2·8	4·24·4	approx quantities	minor changes
26·2·9		CDM	omitted
26·2·10	4·24·3	suspension	unchanged
26·2·11	4·24·5	any impediment	qualification 'save as provided (above)' removed, now covers some matters which were previously specifically listed
26·3		apportionment of EOTs	now covered by 2·28·3
26·4		NSC's	omitted
26·5	4·25	add to Contract Sum	unchanged
26·6	4·26	without prejudice	unchanged
Determination by employer			
27·1	8·2·3	giving notice	reworded to reflect new definitions
27·2·1	8·4·1	default	minor changes
27·2·2	8·4·2, 8·2·2	second notice	unchanged, last sentence goes to 8·2·2
27·2·3	8·4·3	repeat default	minor changes, last sentence goes to 8·2·2
27·2·4	8·2·1	notices	re-grouped
27·3·1	8·1	definition insolvency	re-written completely, presented as a definition (wider scope)
27·3·2	8·5·2	duty to inform	re-written
27·3·3		automatic determination	omitted
27·3·4	8·5·1	right to determine	re-written, automatic termination removed
27·4	8·6	corruption	re-written, simplified – offence under Corruptions Acts only
27·5·1	8·5·3·1, 8·5·3·2	payment	re-written
27·5·2		27·5·2·1 agreement	omitted
27·5·3		interim arrangement	omitted
27·5·4	8·5·3·3	protecting works	re-written – much reduced
27·6	8·7	consequences	re-written – reduced
27·6·1	8·7·1	persons	re-written – reduced
27·6·2·1	8·7·2·3	assignment	re-written – reduced by removing exception
27·6·2·2		direct payment sub-contractors	omitted
27·6·3	8·7·2·1	tools	re-written – much reduced

Table of destinations Continued

clauses	JCT98 (1-4)	SBC05	subject	change
	27·6·3·2	8·7·2·2	CDP	re-written – reduced
	27·6·4·1	8·7·3	applications	large part of clause omitted
	27·6·4·2	8·7·4	account	re-written – reduced
	27·6·5	8·7·4	account	unchanged
	27·6·6	8·7·5	difference	unchanged
	27·7·1	8·8	not completing	re-written – reduced
	27·7·2		employer default	omitted
	27·8	8·3·1	without prejudice	re-grouped

Determination by contractor

	28·1	8·2·3	giving notice	reworded to reflect new definitions
	28·2·1	8·9·1	default	minor changes
	28·2·2	8·9·2	default	re-written – much reduced – general default/ impediment replaces list of specific defaults period of suspension 2 months if none stated
	28·2·3	8·9·3	second notice	unchanged
	28·2·4	8·9·4	repeat default	re-written – reduced
	28·2·5	8·2·1	notices	re-grouped
	28·3·1	8·10·1	definition insolvency	re-written completely, presented as a definition
	28·3·2	8·10·2	duty to inform	re-written
	28·3·3	8·10·1, 8·10·3	right to determine	re-written
	28·4	8·12·1	consequences	consequences of determination by contractor or either party dealt with together – 28·4 rewritten and reduced
	28·4·1	8·12·2	remove tools	reduced
	28·4·2		retention	omitted – retention accounted for elsewhere
	28·4·3	8·12·3, 8·12·5	account	unchanged
	28·5	8·3·1	without prejudice	re-grouped

Determination by either party

	28A·1·1	8·11·1	grounds	period of suspension 2 months if none stated
	28A·1·2	8·11·2	grounds	reduced by incorporation of Contractor's Persons definition
	28A·1·3	8·2·1	notices	re-grouped
	28A·2	8·12·1	consequences	consequences of termination by contractor or either party dealt with together, contractor to prepare account

Table of destinations Continued

clauses	JCT98 (1-4)	SBC05	subject	change
	28A·3	8·12·2	remove tools	reduced
	28A·4		retention	omitted – retention accounted for elsewhere
	28A·5	8·12·3, 8·12·5	account	unchanged
	28A·6	8·12·4	exception	re-written
	28A·7		NSC	omitted
Works by employer				
	29·1	2·7·1	direct engagement	re-structured, but essentially the same
	29·2	2·7·2	direct engagement	re-structured, but essentially the same
	29·3		direct engagement	no equivalent – covered by 'employer's persons' definition
Certificates and Payment				
	30A	4·7	CIS	minor changes, reference to another clause replaced by CIS
	30·1·1·1	4·9·1	Interim Certificates	1st part 1st sentence only
	30·1·1·1	4·13·1	Interim Certificates	2nd part first sentence – final date for payment
	30·1·1·1	4·13·6	interest	references to 'simple interest' replaced by 'the Interest Rate' (defined?)
	30·1·1·2	4·13·2	withholding	unchanged, except last sentence deleted
	30·1·1·3	4·13·3	notice	unchanged
	30·1·1·4	4·13·4	notice	unchanged except cross reference removed
	30·1·1·5	4·13·5	payment	changed significantly
	30·1·1·6	4·8	advance payment	wording simplified, bond to be provided unless stated otherwise
	30·1·2·1	4·11	interim valuations	footnote reference to formula adjustment, clause 40·2·1, incorporated in main text
	30·1·2·2	4·12	contractor's application	minor changes
	30·1·3	4·9·2	issue of Interim Certificates	minor changes
	30·1·4	4·14	suspension	last sentence omitted
	30·2	4·10, 4·16	amount due	minor changes
	30·2·1	4·16·1	inclusions, subject to retention	unchanged
	30·2·1·1	4·16·1·1	value of work	Price Statement replaced by Schedule 2 quotation, Fluctuations incorporated, reference to Activity Schedule retained and simplifietd

Table of destinations Continued

clauses	JCT98 (1-4)	SBC05	subject	change
	30·2·1·2	4·16·1·2	materials on site	unchanged
	30·2·1·3	4·16·1·3	listed items	simplified
	30·2·1·4		NSC amounts	omitted
	30·2·1·5		NSC profit	omitted
	30·2·2	4·16·2	inclusions, not subject to retention	unchanged
	30·2·2·1	4·16·2·1	clause 3 amounts	unchanged
	30·2·2·2	4·16·2·2	insurance amounts	unchanged
	30·2·2·3		NSC amounts	omitted
	30·2·2·4	4·16·2·3	fluctuations	unchanged
	30·2·2·5		NSC amounts	omitted
	30·2·3·1	4·16·3	deductions	clause split into 2 parts
	30·2·3·2		NSC deductions	omitted
	30·3	4·17	listed items	first sentence omitted
	30·3·1	4·17·2·1, 4·17·4	property vesting	wording simplified
	30·3·2	4·17·2, 4·17·5	listed items	wording simplified
	30·3·3	4·17·1	listed items	unchanged
	30·3·4	4·17·3	identification	re-drafted
	30·3·5	4·17·2·2	insurance	wording simplified
	30·4·1	4·20	retention rules	wording simplified
	30·4·1·1	4·20·1	rate	re-drafted, much shorter, rate changed to 3%
	30·4·1·2	4·20·2	retention	minor changes
	30·4·1·3	4·20·3	half retention	minor changes
	30·4·2		definition	omitted
	30·4A	4·19	bond in lieu retention	minor changes, surety to be approved by employer
	30·5	4·18	retention	
	30·5·1	4·18·1	fiduciary interest	references to NSC removed
	30·5·2	4·18·2	statement	second part (30·5·2·2) omitted
	30·5·3	4·18·3	bank account	minor changes
	30·5·4	4·18·4	informing contractor	minor changes
	30·6·1	4·5	final adjustment	minor changes
	30·6·2	4·3·1	contract sum adjustments	re-worded to reflect simplified provisions for valuing variations

Table of destinations Continued

clauses	JCT98 (1-4)	SBC05	subject	change
	30·6·2·1–5	4·3·2	deductions	minor changes
	30·6·2·6–9			omitted
	30·6·2·10–16	4·3·3	inclusions	minor changes
	30·6·2·17		price statements	omitted
	30·7		NSC amounts	omitted
	30·8·1	4·15·1	Final Certificate	minor changes
	30·8·1·1	4·15·2	Final Certificate	re-structured, reference to stating how amount calculated moved to end of clause
	30·8·2	4·15·3	first notice	minor changes
	30·8·3	4·15·4	second notice	unchanged
	30·8·4	4·15·5	payment	unchanged
	30·8·5	4·15·6	interest	references to 'simple interest' replaced by 'the Interest Rate' (defined?)
	30·8·6	4·15·7	balance	unchanged
	30·9·1	1·10·1	conclusiveness	minor changes
	30·9·2	1·10·2	effect of proceedings	minor rewording
	30·9·3	1·10·3	effect of proceedings	unchanged
	30·9·4	1·10·4	effect on decisions	minor changes
	30·10	1·11	effect of certificates	minor changes
	31	4·7	CIS	omitted
	32		not used	
	33		not used	
Antiquities				
	34·1	3·22	antiquities	unchanged
	34·2	3·23	antiquities	last sentence removed
	34·3	3·24	antiquities/loss expense	section on extension of time removed
35: Nominated sub-contractors				omitted
36: Nominated suppliers				omitted
Fluctuations				
	37·1	4·21	options	reworded, footnote incorporated
	37·2		default option	omitted. Now in contract particulars
	37·3	4·22	13A etc.	reworded

Table of destinations Continued

clauses	JCT98 (1-4)	SBC05	subject	change
	38	Schedule 7:A	contributions, etc.	minor changes
	39	Schedule 7:B	labour etc.	minor changes
	40	Schedule 7:C	formula	minor changes
	Dispute resolution			
	41, footnote [uu]	9·1	mediation	footnote moved to a clause, refers to 'Guidance Note'
	Adjudication			
	41A	9·2	adjudication	JCT provisions replaced by the Scheme, with qualifications relating to who will be the adjudicator, and 3·18·4 instructions (further tests)
	Arbitration			
	41B	9·3–9·8	rules	unchanged, except that 41B·6 moved to 9·3
	Legal proceedings			
	41C		legal proceedings	omitted
	Performance Specified Work			
	42		PSW	omitted
	Code of practice referred to in clause 8·4·4			
	CP 1 and 2	Schedule 4	further testing	unchanged
	Appendix			omitted
	Annex 1	Schedule 6	Part 1: advance payment bond	unchanged
		Schedule 6	Part 2: off-site bond	unchanged
	Supplemental provs		VAT	omitted
	Annex 2		EDI	omitted
	Annex 3	Schedule 6	Part 3: retention bond	unchanged

Appendix B: *Clause comparison table: Origins*

Table of origins		
	SBC05	**JCT98 (1-4)**
Recitals	first	first
	second	second
	third	third
	fourth	fourth
	fifth	sixth
	sixth	third (CDPS/SC)
	seventh	second (CDPS/SC)
	eighth	third (CDPS/SC)
	ninth	fourth (CDPS/SC)
	tenth	tenth (CDPS/SC)
Articles	Article 1	Article 1
	Article 2	Article 2
	Article 3	Article 3
	Article 4	Article 4
	Article 5	Article 6.1
	Article 6	Article 6.2
	Article 7	Article 5
	Article 8	Article 7A
	Article 9	Article 7B
	Section 1: Definitions and interpretation	
Clauses	1·1	1·3
	1·2	1·1
	1·3	1·2, 2·2·1
	1·4	
	1·5	1·8
	1·6	1·12
	1·7	1·7
	1·8	1·11
	1·9	5·8
	1·10·1	30·9·1
	1·10·2	30·9·2
	1·10·3	30·9·3
	1·10·4	30·9·4
	1·11	30·10
	1·12	1·10
	Section 2: Carrying out the Works	
	2·1	2·1, 6·1·1, 8·1·3
	2·2	2·1·2

Table of origins Continued

clauses	SBC05	JCT98 (1-4)
	2·3·1	8·1·1, 8·1·4
	2·3·2	8·1·2
	2·1, 8·1·2	2·1·3
	2·3·4	8·2·1
	2·3·5	
	2·4	23·1·1
	2·4·1	23·3·1
	2·4·2	23·3·1
	2·5	23·1·2
	2·6·1	23·3·2
	2·6·2	23·3·3
	2·7·1	29·1
	2·7·2	29·2
	2·8·1	5·1
	2·8·2	5·2
	2·8·3	5·5
	2·8·4	5·7
	2·9·1	5·3·2
	2·9·1·1	5·3·1·1
	2·9·1·2	5·3·1·2
	2·9·2	2·6·1
	2·9·3	2·6·2
	2·10	7
	2·11	5·4·1
	2·12	5·4·2
	2·13·1	2·2·2·1
	2·13·2	
	2·14·1	2·2·2·2
	2·14·2	
	2·14·3	2·2·2·2
	2·14·4	2·2·2·3
	2·15	2·3
	2·16	2·5·1
	2·17·1	6·1·2, 6·1·8
	2·17·2	6·1·3
	2·17·3	6·1·5
	2·18	6·1·4, 6·1·9
	2·19·1	2·7·1
	2·19·2	2·7·2
	2·19·3	2·7·3
	2·20	2·10
	2·21	6·2
	2·22	9·1
	2·23	9·2
	2·24	16·1
	2·25	16·2

Table of origins Continued

clauses	SBC05	JCT98 (1-4)
	2·26	25·1
	2·27·1	25·2·1
	2·27·2	25·2·2
	2·27·3	25·2·3
	2·28·1	25·3·1
	2·28·1·1	25·3·1·1
	2·28·1·2	25·3·1·2
	2·28·2	25·3·1
	2·28·3·1	25·3·1·3
	2·28·3·2	25·3·1·4
	2·28·4	25·3·2
	2·28·5	25·3·3
	2·28·6	25·3·4, 25·3·6
	2·29·1	25·4·5·1
	2·29·2	25·4·5·1, 25·4·5·2
	2·29·3	25·4·13
	2·29·4	25·4·14
	2·29·5	25·4·18
	2·29·6	25·4·19
	2·29·7	25·4·11
	2·29·8	25·4·1
	2·29·9	25·4·3
	2·29·10	25·4·16
	2·29·11	25·4·4
	2·29·12	25·4·9
	2·29·13	25·4·2
	2·30	17·1
	2·31	24·1
	2·32·1	24·2·1
	2·32·2	24·2·1
	2·32·3	24·2·2
	2·32·4	24·2·3
	2·33	18·1
	2·34	18·1·1
	2·35	18·1·2
	2·36	18·1·3
	2·37	18·1·4
	2·38·1	17·2
	2·38·2	17·3
	2·39	17·4
	2·40	5·10
	2·41	
	Section 3: Control of the Works	
	3·1	11
	3·2	10
	3·3	1·9

Table of origins Continued

clauses	SBC05	JCT98 (1-4)
	3·4	12
	3·5	article 3, article 4
	3·6	1·5
	3·7·1	19·2·2
	3·7·2	19·2·3
	3·7·3	6·3
	3·8·1	19·3·1
	3·8·2	19·3·2
	3·8·2	19·3·2
	3·8·4	19·3·3
	3·9	19·4
	3·9·1	19·4·1
	3·9·2·1	19·4·2·1
	3·9·2·1·1	19·4·2·2
	3·9·2·1·2	19·4·2·3
	3·9·2·2	
	3·9·2·3	19·4·3
	3·9·2·4	
	3·9·2·5	19·4·2·4
	3·9·3	
	3·10·1	4·1·1
	3·10·2	4·1·1
	3·10·3	2·8
	3·11	4·1·2
	3·12	4·3
	3·13	4·2
	3·14	13·2
	3·15	23·2
	3·16	13·3·1
	3·17	8·3
	3·18·1	8·4·1
	3·18·2	8·4·2
	3·18·3	8·4·3
	3·18·4	8·4·4
	3·19	8·5
	3·20	8·2·2
	3·21	8·6
	3·22	34·1
	3·23	34·2
	3·24	34·3
	3·25·1	6A·1
	3·25·2	6A·2
	3·25·3	6A·4
	3·26	1·6, 6A·3
	Section 4: payment	
	4·1	14·1

Table of origins Continued

clauses	SBC05	JCT98 (1-4)
	4·2	14·2
	4·3·1	30·6·2
	4·3·2	30·6·2·1–5
	4·3·3	30·6·2·10–16
	4·4	3
	4·5	30·6·1
	4·6·1	15·2
	4·6·2	15·3
	4·7	30A
	4·8	30·1·1·6
	4·9·1	30·1·1·1
	4·9·2	30·1·3
	4·10	30·2
	4·11	30·1·2·1
	4·12	30·1·2·2
	4·13·1	30·1·1·1
	4·13·2	30·1·1·2
	4·13·3	30·1·1·3
	4·13·4	30·1·1·4
	4·13·5	30·1·1·5
	4·13·6	30·1·1·1
	4·14	30·1·4
	4·15·1	30·8·1
	4·15·2	30·8·1·1
	4·15·3	30·8·2
	4·15·4	30·8·3
	4·15·5	30·8·4
	4·15·6	30·8·5
	4·15·7	30·8·6
	4·16	30·2
	4·16·1	30·2·1
	4·16·1·1	30·2·1·1
	4·16·1·2	30·2·1·2
	4·16·1·3	30·2·1·3
	4·16·2	30·2·2
	4·16·2·1	30·2·2·1
	4·16·2·2	30·2·2·2
	4·16·2·3	30·2·2·4
	4·16·3	30·2·3·1
	4·17	30·3
	4·17·1	30·3·3
	4·17·2	30·3·2
	4·17·2·1	30·3·1
	4·17·2·2	30·3·5
	4·17·3	30·3·4
	4·17·4	30·3·1

Table of origins Continued

clauses	SBC05	JCT98 (1-4)
	4·17·5	30·3·2
	4·18	30·5
	4·18·1	30·5·1
	4·18·2	30·5·2
	4·18·3	30·5·3
	4·18·4	30·5·4
	4·19	30·4A
	4·20	30·4·1
	4·20·1	30·4·1·1
	4·20·2	30·4·1·2
	4·20·3	30·4·1·3
	4·21	37·1
	4·22	37·3
	4·23	26·1
	4·24	26·2
	4·24·1	26·2·7
	4·24·2·1	26·2·5
	4·24·2·1	26·2·7
	4·24·2·2	26·2·2
	4·24·2·3	26·2·3
	4·24·3	26·2·10
	4·24·4	26·2·8
	4·24·5	26·2·11
	4·25	26·5
	4·26	26·6
	Section 5: Variations	
	5·1·1	13·1·1
	5·1·2	13·1·2
	5·2·1	13·4·1·3
	5·2·1	13·4·1·1
	5·2·2	13·4·1·1
	5·3	13A
	5·3·1	13A·1·1
	5·3·1	13·2·3
	5·3·2	13·2·3
	5·3·3	13A·8
	5·4	13·6
	5·5	13·7
	5·6·1	13·5·1
	5·6·1·1	13·5·1·1
	5·6·1·2	13·5·1·2
	5·6·1·3	13·5·1·3
	5·6·1·3	13·5·1·4
	5·6·1·4	13·5·1·5
	5·6·2	13·5·2
	5·6·3	13·5·3

Table of origins Continued

clauses	**SBC05**	**JCT98 (1-4)**
	5·7	13·5·4
	5·8	13·8
	5·9	13·5·5
	5·10	13·5·7
	Section 6: Injury, damage and insurance	
	6·1	20·1
	6·2	20·2
	6·3·1	20·3·1
	6·3·2	
	6·3·3	20·3·2
	6·4·1	21·1·1
	6·4·2	21·1·2
	6·4·3	21·1·3
	6·5·1	21·2·1
	6·5·2	21·2·2
	6·5·3	21·2·3
	6·6	21·3
	6·7	22·1
	6·8	22·2
	6·9	22·3
	6·10·1	22A·5·1, 22B·4·1, 22C·5·1
	6·10·2	22A·5·2, 22B·4·2, 22C·5·2
	6·10·3	22A·5·2, 22B·4·2, 22C·5·2
	6·10·4	22A·5·3, 22B·4·3, 22C·5·3
	6·11	
	6·12	
	6·13	22FC·1
	6·14	22FC·2
	6·15	22FC·3
	6·16	22FC·5
	Section 7: Assignment, third party rights and collateral warranties	
	7·1	19·1·1
	7·2	19·1·2
	7·3	
	7·4	
	7·5	
	7A–7F	
	Section 8: Termination	
	8·1	27·3·1
	8·2·1	27·2·4, 28·2·5, 28A·1·3
	8·2·2	27·2·2
	8·2·3	27·1, 28·1
	8·3·1	27·8, 28·5
	8·3·2	
	8·4·1	27·2·1
	8·4·2	27·2·2

Table of origins Continued

clauses	SBC05	JCT98 (1-4)
	8·4·3	27·2·3
	8·5·1	27·3·4
	8·5·2	27·3·2
	8·5·3·1	27·5·1
	8·5·3·2	27·5·1
	8·5·3·3	27·5·4
	8·6	27·4
	8·7	27·6
	8·7·1	27·6·1
	8·7·2·1	27·6·3
	8·7·2·2	27·6·3·2
	8·7·2·3	27·6·2·1
	8·7·3	27·6·4·1
	8·7·4	27·6·4·2, 27·6·5
	8·7·5	27·6·6
	8·8	27·7·1
	8·9·1	28·2·1
	8·9·2	28·2·2
	8·9·3	28·2·3
	8·9·4	28·2·4
	8·10·1	28·3·1, 28·3·3
	8·10·2	28·3·2
	8·10·3	28·3·3
	8·11·1	28A·1·1
	8·11·2	28A·1·2
	8·12·1	28·4, 28A·2
	8·12·2	28·4·1, 28A·3
	8·12·3	28·4·3, 28A·5
	8·12·4	28A·6
	8·12·5	28·4·3, 28A·5·5
	Section 9: Settlement of disputes	
	9·1	41, footnote [uu]
	9·2	41A
	9·3–9·8	41B
	Schedule 1: Design submission Procedure	
	S·1 1–8	
	Schedule 2: Schedule 2 Quotation	
	S·2 1–7	13A
	Schedule 3: Insurance options	
	S·3 A·1	22A·1
	S·3 A·2	22A·2
	S·3 A·3	22A·3
	S·3 A·4	22A·4
	S·3 A·5	22A·5·4
	S·3 B·1	22B·1
	S·3 B·2	22B·2

Table of origins Continued

clauses	**SBC05**	**JCT98 (1-4)**
	S·3 B·3	22B·3
	S·3 C·1	22C·1
	S·3 C·2	22C·2
	S·3 C·3	22C·3
	S·3 C·4	22C·4
	Schedule 4	CP 1 and 2
	Schedule 5	
	Schedule 6	Annex 1
	Schedule 6	Annex 1
	Schedule 6	Annex 3
	Schedule 7	

Appendix C: *Activity Schedule*

Example of priced Activity Schedule (see item 13)

A	Preliminaries – see breakdown	170,000.00
B	Demolition	40,000.00
C	Substructure including ground floor slab	178,000.00
D	Structural frame	265,000.00
E	Upper floor and staircase structures	112,000.00
F	Roof structure and coverings	58,000.00
G	External walls	203,000.00
H	Windows and external doors	102,000.00
J	Internal walls and doors therein	75,000.00
K	Partitioning and doors therein	45,000.00
L	Plasterwork	21,000.00
M	Screeds	33,000.00
N	Suspended ceilings	53,000.00
P	Wall tiling	16,000.00
Q	Floor tiling	11,000.00
R	Other floor finishings	64,000.00
S	Metalwork	30,000.00
T	Fittings and fixtures	35,000.00
U	Decorations	12,000.00
V	Sanitary installation and fittings	48,000.00
W	Rainwater installation	8,500.00
X	Mechanical services – see breakdown	297,000.00
Y	Electrical services – see breakdown	222,000.00
Z	Underground drainage	16,500.00
AA	External works	75,000.00
		£2,190,000.00

Note that the above excludes P.C. Sums and profit thereon, Provisional Sums and Provisional Quantities.

Priced Activity Schedule

Example breakdowns

Preliminaries

A	Site accommodation (offices, stores, toilets, etc.)	6,000.00
B	Services (power, lighting, telephones, small plant, rubbish disposal, etc.)	11,000.00
C	Mechanical plant	8,500.00
D	Temporary works (roads, walkways, scaffolding, hoardings, etc.)	35,000.00
E	Management and staff	92,500.00
F	The Contract, etc. (insurance premiums, cost of providing Bond, etc.)	7,000.00
		£170,000.00

Mechanical services

A	Heating installation	235,000.00
B	Hot water installation	6,000.00
C	Cold water installation including rising main and tank	14,000.00
D	Gas installation	2,500.00
E	Mechanical ventilation	25,000.00
F	Controls and wiring	9,000.00
G	Testing and commissioning	3,000.00
H	As installed record drawings	1,500.00
J	Operation and maintenance manuals	1,000.00
		£297,000.00

Electrical services

A	Distribution boards, switchgear and sub mains cabling	28,000.00
B	Power installation	72,000.00
C	Lighting installation including fittings	92,000.00
D	Fire alarm system	14,500.00
E	Security system	6,000.00
F	Earthing and bonding	5,000.00
G	Testing and certification	2,500.00
H	As installed record drawings	1,250.00
J	Operation and maintenance manuals	750.00
		£222,000.00

References

Publications

Furst, S. and Ramsay, V. *Keating on Building Contracts*, Sweet & Maxwell (2001).
Hyams, D. *Construction Companion to Briefing*, RIBA Publications (2001).

Cases

Alfred McAlpine Homes North Ltd v *Property and Land Contractors Ltd* (1995) 76 BLR 59
Archivent Sales & Developments Ltd v *Strathclyde Regional Council* (1984) 27 BLR 98 (Court of
        Session, Outer House)
B R Cantrell (2) E P Cantrell v *Wright and Fuller Ltd* [2003] BLR 412
B Mullan & Sons Contractors v *John Ross* (1996) 86 BLR 1
Balfour Beatty Building Ltd v *Chestermont Properties Ltd* (1993) 62 BLR 1
Bath and North East Somerset DC v *Mowlem plc* [2004] BLR 153
British Telecommunications plc v *James Thompson & Sons (Engineers) Ltd* [1999] BLR 35 (HL)
C M Pillings & Co. Ltd v *Kent Investments Ltd* (1985) 30 BLR 80 (CA)
City of Westminster v *J Jarvis & Sons Ltd* (1970) 7 BLR 64 (HL)
Co-operative Insurance Society v *Henry Boot Scotland and others* [2002] CILL 1932
Cooperative Retail Services Limited v *Taylor Young Partnerships* [2002] BLR 272
Croudace Ltd v *The London Borough of Lambeth* (1986) 33 BLR 25 (CA)
Crown Estate Commissioners v *John Mowlem & Co. Ltd* (1994) 70 BLR 1 (CA)
Dawber Williams Roofing Ltd v *Humberside County Council* (1979) 14 BLR 70
Department of Environment for Northern Ireland v *Farrans (Construction) Ltd* (1981) 19 BLR 1
        (NI)
F G Minter Ltd v *Welsh Health Technical Services Organisation* (1980) 13 BLR 1 (CA)
Gilbert-Ash (Northern) Ltd v *Modern Engineering (Bristol) Ltd* (1974) 1 BLR 73 (HL)
Glenlion Construction Ltd v *The Guinness Trust* (1987) 39 BLR 89
Gloucester County Council v *Richardson* [1968] 1 AC 480 (CA)
Greater London Council v *The Cleveland Bridge and Engineering Co. Ltd* (1986) 34 BLR 50 (CA)
H Fairweather & Co. Ltd v *London Borough of Wandsworth* (1987) 39 BLR 106
H W Neville (Sunblest) Ltd v *William Press & Son Ltd* (1981) 20 BLR 78
Holland Hannen & Cubitts (Northern) Ltd v *Welsh Health Technical Services Organisation* (1985)
        35 BLR 1 (CA)
Henry Boot Construction Ltd v *Alstom Combined Cycles* [2000] BLR 247
Impresa Castelli SpA v *Cola Holdings Ltd* (2002) CLJ 45
J F Finnegan Ltd v *Community Housing Association Ltd* (1995) 75 BLR 22 (CA)
J F Finnegan Ltd v *Ford Sellar Morris Developments Ltd* (1991) 53 BLR 38
Kensington and Chelsea and Westminster Health Authority v *Wettern Composites* (1984) 31 BLR 57
Kruger Tissue (Industrial) Ltd v *Frank Galliers Ltd* (1998) 57 ConLR 1
Leedsford Ltd v *The Lord Mayor, Alderman and Citizens of the City of Bradford* (1956) 24 BLR 45
        (CA)
London Borough of Barking & Dagenham v *Terrapin Construction Ltd* [2000] BLR 479
London Borough of Hounslow v *Twickenham Garden Developments* (1970) 7 BLR 81
London Borough of Merton v *Stanley Hugh Leach Ltd* (1985) 32 BLR 51 (Chancery Division)

Lubenham Fidelities and Investments Co. Ltd v *South Pembrokeshire District Council* (1986) 33 BLR 39 (CA)

Mac-Jordan Construction Ltd v *Brookmount Erostin Ltd* (1991) 56 BLR 1

Michael Salliss & Co. Ltd v *Calil and William F. Newman & Associates* (1987) 13 ConLR 69

MOD v *Scott Wilson Kirkpatrick and Partners* [2000] BLR (CA)

Pacific Associates Inc v *Baxter* (1988) 44 BLR 33 (CA)

Peak Construction (Liverpool) Ltd v *McKinney Foundations Ltd* (1970) 1 BLR 111 (CA)

Pearce and High v *John P. Baxter and Mrs A. S. Baxter* [1999] BLR 101 (CA)

Penwith District Council v *V P Developments Ltd* 21 May 1999, unreported

Percy Bilton Ltd v *The Greater London Council* (1981) 20 BLR 1

Plant Construction v *Clive Adams Associates and JMH Construction Services* [2000] BLR 158 (CA)

R Burden Ltd v *Swansea Corporation* [1957] 3 All ER 243 (HL)

Rupert Morgan Building Services (CCC) Ltd v *David Jervis and Hamlet Jervis* [2003] EWCA Cir 1583 CA

Scottish Special Housing Association v *Wimpey Construction UK Ltd* (1986) 34 BLR 1

Scottish and Newcastle plc v *G. D. Construction (St Albans) Ltd* [2003] BLR 131

Skanska Construction (Regions) Ltd v *Anglo-Amsterdam Corporation Ltd* (2002) 84 ConLR 100

Sutcliffe v *Chippendale & Edmondson* (1971) 18 BLR 149

Sutcliffe v *Thackrah* (1974) 4 BLR 16 (CA)

Temloc v *Errill Properties* (1988) 39 BLR 30 (CA)

Townsend v *Stone Toms & Partners* (1984) 27 BLR 26 (CA)

Viking Grain Storage v *T H White Installations* (1985) 33 BLR 103

Wates Construction (London) Ltd v *Franthom Property Ltd* (1991) 53 BLR 23 (CA)

Wates Construction (South) Ltd v *Bredero Fleet Ltd* (1963) 63 BLR 128

West Faulkner Associates v *London Borough of Newham* (1992) 61 BLR 81

Whittal Builders Co. Ltd v *Chester-le-Street District Council* (1987) 40 BLR 82

Clause Index *by paragraph number*

Subject Index *by paragraph number*